Abraham Jacobi

The Intestinal Diseases of Infancy and Childhood

Physiology, hygiene, pathology and therapeutics

Abraham Jacobi

The Intestinal Diseases of Infancy and Childhood
Physiology, hygiene, pathology and therapeutics

ISBN/EAN: 9783337822408

Printed in Europe, USA, Canada, Australia, Japan

Cover: Foto ©berggeist007 / pixelio.de

More available books at **www.hansebooks.com**

THE FAIRCHILD PREPARATIONS
– OF –
THE PURE DIGESTIVE FERMENTS,
Active, Permanent and Reliable.

TRYPSIN
(FAIRCHILD)

Especially Prepared as a Solvent for Diphtheritic Membrane.

PEPTONISING TUBES.
(FAIRCHILD).

For the preparation of PEPTO-NIZED MILK and other predigested food for the sick.

PEPSINE IN SCALES.
(FAIRCHILD).

The most active, permanent and reliable pepsine made in the World.

ESSENCE OF PEPSINE
(FAIRCHILD).

For administration where a fluid and agreeable form of pepsine is desired, and for the preparation of Junket and Whey.

EXTRACTUM PANCREATIS.
(FAIRCHILD).

Containing all the digestive ferments of the Pancreas.

PEPTOGENIC MILK POWDER
(FAIRCHILD).

For the modification of cows' milk to the standard of Normal Mother's Milk.

PEPSINE IN POWDER.
(FAIRCHILD).

Prepared from the scales without the admixture of any other substances, to facilitate dispensing and the preparation of saccharated pepsine.

DIASTASIC ESSENCE OF PANCREAS.
(FAIRCHILD).

For the digestion of starchy foods.

FAIRCHILD BROS. & FOSTER,
82 AND 84 FULTON ST., NEW YORK.

DR. L. S. DE FOREST,
NEW HAVEN, CONN.

—THE—

INTESTINAL DISEASES OF INFANCY AND CHILDHOOD.

Physiology, Hygiene, Pathology and Therapeutics.

— BY —

A. JACOBI, M. D.,

Ex-President of the New York Academy of Medicine; Clinical Professor of Diseases of Children in the College of Physicians and Surgeons, New York, etc.

VOL. I.

SECOND EDITION.

1890.
GEORGE S. DAVIS.
DETROIT, MICH.

Copyrighted by
GEORGE S. DAVIS
1890.

TABLE OF CONTENTS.

VOL. 1.

	PAGE
ALTERATIONS IN BREAST-MILK	1
DIET OF WET-NURSES	13
SUBSTITUTE FOR BREAST-MILK	16
MILK OF ONE COW	16
RAW OR BOILED MILK	17
CONDENSED MILK	19
OTHER MILKS	20
FAT	23
CASEIN	24
ADDITION OF WATER TO COW'S MILK	31
ADMIXTURE OF SALT	35
ANIMAL ADMIXTURES	41
OTHER SUBSTITUTES	44
SALIVA AND STARCH	47
HOW TO FEED	67
NOURISHMENT SUCCEEDING THE PERIOD OF INFANCY,	73
THE MOUTH	77
THE STOMACH	82
GASTRIC DIGESTION	95
GASTRO-INTESTINAL DISEASES	102
DYSPEPSIA	103
POLYPHAGIA—BULIMIA	106
VOMITING	107
ACUTE GASTRIC CATARRH	109
CORROSIVE GASTRITIS	114
DIPHTHERITIC GASTRITIS	116
CHRONIC GASTRIC CATARRH	116
HÆMORRHAGE FROM STOMACH	127
ULCERATION OF STOMACH	127
TUMORS OF STOMACH	137

VI.
VOL. II.

	PAGE
Intestinal Digestion	139
Meconium and Fæces	142
Liver	145
Pancreas	146
Symptomatology of Intestinal Diseases	153
Colic	153
Constipation	154
Predisposition to Diarrhœas	159
Fat Diarrhœa	161
Acute Intestinal Catarrh	162
Chronic Intestinal Catarrh	171
Follicular Enteritis	174
Membranous Enteritis	176
Diphtheritic or Croupous Enteritis	178
Ulceration of Intestines	179
Tuberculosis of Intestines and Mesenteric Glands	194
Perityphlytis	198
Paratyphlitis	204
Invagination—Intussusception	206
Parasites—Worms—Helminthiasis	213
Intestinal Malformations	222
Hernias—Ruptures	233
Inflammation of Rectum—Proktitis	243
Dysentery	244
Polypus of Rectum	250
Prolapse of Rectum	254
Fissure of Anus	259

INTRODUCTION.

Of all the fatal affections that occur in the first year of life, forty per cent. are diseases of the digestive, and twenty per cent. diseases of the respiratory organs.

In the second year the main causes of death change entirely, for of forty-five deaths from the two causes, but nine are due to diseases of the digestive, and thirty-six to affections of the respiratory organs.

Thus, in the first year the stomach and intestines, and in the second the bronchi and lungs, are the sources of high death-rate. In New York City, however, we meet with a high rate of mortality from digestive disorders even in children of more than one year of age, and the second summer is therefore regarded with awe and fear amounting to superstition. In fact, public opinion looks for a higher rate of mortality in the second than in the first summer. The fallacy of this assumption can be easily corrected by statistical reports, and the high mortality itself should be easily reduced by such parents as can become convinced that it is external causes which kill their children, and not the natural course of development. The second summer is a period of danger, in part only, because of the heat, but mostly because of errors in feeding. Conscientious and intelligent families, in good circumstances, are not liable to lose their infants in the second summer.

Mortality diminishes with every day of advancing life: every additional hour improves the baby's chances for preservation. Almost one-half of the infants who die before the end of the first year, do so before they are one month old. The causes of disease are the more active, the earlier they are brought to bear upon the young with their defective vitality. Two grave conclusions are to be drawn from this fact.

First. That diminution of early mortality depends upon avoiding diseases of the digestive organs by insisting on normal alimentation. This is particularly important in the first few months. While it has been shown that breast milk lowers the rate of infant mortality through the entire first year, it does so much more in the first few months; thus, although infants may not be fed on breast milk through the whole of the normal period for nursing, very great gain is accomplished by insisting that they shall be nursed for at least a limited time, if only a few months. There are few mothers but are capable of nursing during that brief period, and all contribute to the illness or death of their babes by refusing to nurse them through at least the first dangerous weeks.

Second. The hygienic rules for infants concern the digestive organs mainly—so much so, that infant hygiene, and the hygiene of the digestive organs in infants, appear to be nearly identical.

These are the reasons which induced me to make the digestive organs of infants and children the subject of this brief manual, though in so doing, I am aware of the frequent failures of attempts made at saving those who have fallen ill, and of the greater possibility to prevent disease than to cure it. Moreover, it appears to me no thankless task to look upon the pathological changes which become the subjects of our therapeutical endeavors, as the mere results of anomalies of the physiological functions. Thus I propose to treat first of the physiology of the digestion of childhood, and the main points of its hygiene, as their lessons render the morbid changes more intelligible.

Some of my readers may know that I have written before on these and similar subjects. I refer only to my "Infant Diet," 2d edition, 1374; my contribution to A. Buck's "Treatise on Hygiene and Public Health," vol. i, New York, 1879; and to C. Gerhardt's "Manual of the Diseases of Children," first volume, 2d edition, 1882, and occasional articles in medical journals. Besides, my practical experience, extending over more than a third of a century, may justify this attempt at appearing before the medical public with expositions, which cannot be considered trite and superfluous as long as the mortality from digestive disorders amongst infants and children is as appalling as it is known to be.

A. JACOBI, M. D.

110 West 34th St.. New York, October, 1887.

ALTERATIONS OF BREAST MILK.

In many of the digestive disorders of infants the best preventive, and often curative, aid is the breast-milk of mother or wet-nurse. This is an axiom, and indisputable law of nature, as long as the circumstances of the case are favorable. In view of the great mortality in the first few months, breast-milk is the one and indispensable food for those of that age. It is true that a babe may be taken sick with intestinal disease in spite of being nursed at the breast, for there are *many* causes of disease; it may, indeed, occur that babes are taken sick *because* of being at the breast. And it is those cases that both mothers and physicians should be well acquainted with. Sometimes it is not the breast-milk which is at fault in the beginning, but the faulty use to which it is put. Many babes suffer intensely because they are not limited to intervals of from 2 to 5 hours, as required by either age or constitution. In such cases, by too frequent feeding, both the milk of the mother and the digestion of the infant are impaired. Here regularity is the sole indication. Sometimes, though fortunately in few cases only, there appears to exist an unexplained idiosyncrasy on the part of the baby, who cannot thrive on the milk of the mother, but succeeds in so doing after a change of food. In many cases, however, there are demonstrable dangers in the breast-milk of either mother or nurse; there may be an undue percentage of fat, of casein, of salts, of sugar, or even accidental admixtures. These may occur in the secreting organ (thus blood may be found in the milk) or be traceable to the circulating blood of the whole system: of the latter they may be the very constituents, or foreign bodies floating in it. They can be classed as either morbid dispositions or as actual admixtures. Women suffering from constitutional syphilis, chronic consumption, anæmia, extensive rhachitis, or severe nervous derangement of hysterical or other origin—those suffering from care and hard work, and those who are compelled to take a great deal of medicine—will serve their infants best by not nursing at all.

In regard to the influence of medicines, opinion is divided. It has been claimed that milk, being a secretion of a gland and not a transudation from the blood, could not contain foreign material to any great extent, but this is true only so far as an absolutely healthy woman and normal milk are concerned. The first period of lactation yields colostrum, not normal milk, and very often the latter is changed into a colostral condition, similar to that after birth, containing different-shaped fat globules, more sugar, and soluble albumin—in fact, real blood serum. This may take place in any or every case of impaired health, and the more serum contained in any milk, the easier is the admixture of soluble substances circulating in the blood. As I formulated it some years ago, milk secreted from insufficient mammæ—by a woman devoid of health and vigor, by an old woman, by a very young woman, by an anæmic woman, by a convalescent woman who has consumed a large portion of her albumin (be it circulating or tissue albumin), by a neurotic woman with frequent disturbances of the circulation—milk, in fact, which is not the normal transformation of the elements of the mammary glands, but instead consists of more or less transuded serum, is apt to be impregnated with elements circulating in the blood. The indications on the one hand for permission to nurse, on the other for the administration of medicines to a nursing woman, require, therefore, greater strictness than is usually conceded. At all events, the good results obtained by artificial feeding, in many cases of ailment on the part of infants, are often more than mere coincidences.

Changes in animal and in woman's milk are frequent, in consequence of nursing, disease, and organic and inorganic materials absorbed, and such occurrence may take place at times when exact analyses are out of the question.

One of the many pretty stories of the younger Pliny is as follows: Lysippe, Iphinœ and Iphianassa, the three daughters of Proteus, King of Argos, were poisoned by the milk of goats that had fed on hellebore. Becoming insane, they roamed the country, seducing and abducting other young girls. Finally they were cured by Melampus and his brother Bias, who married them, and the other victims by vigorous young men who hunted them up.

In Gerhardt's Handb. d. Kinderk. Vol. II, I have collected a large amount of material, partly from reliable veterinary literature, showing the influence of illness, different foods, and drugs upon the milk of animals.

Zukowski observed that wet-nurses when tired and hungry, secreted milk that was not nourishing. At the foundling-hospital in Moscow, the percentage of fat in the milk of wet-nurses when first admitted was from 1.8 to 3.0 per cent.; among those who had been in the institution a short time, it was 3.2 to 4.0 per cent. Fasting exerted great influence upon the milk, especially as regards the fatty element, and many nurslings were liable to be ill at such times. Upon the first day of fast, fat usually decreased to 0.88 per cent., but rose to 3.4 per cent. by gradual habituation to retrenchment in diet, and probably by the appropriation of an extra quantity of albumin from the general system which satisfied the demand of the milk glands. The instrumentalities most rapid in their effect upon the milk, are those derived through the nervous system, and their action is upon an organ in which, when functionally active, rapid changes occur.

Firmin (Bull. Thérap., 1886; Schmidt's Jahrbücher, 1875, No. 8) reports the case of a child six months of age, attacked with urticaria, fainting, vomiting, and offensive diarrhœa, produced by milk after the mother had partaken of oysters, crabs, cod-fish, and shad. R. Monti (Schmidt's Jahrb. 173, p. 160) observed, when the right arm of a nursing woman (whose breast of the same side was functionally incapacitated by mastitis), was treated locally with ammonium and camphor, there was marked decrease of the secretion of milk in the gland of the healthy side. The passage of coloring matter into the tissues within a short space of time is a well-known possibility, and according to Mosler, Schauenstein and Späth, milk will become yellow through the influence of marsh turnips, caltha palustris, saffron, and rhubarb; red after the ingestion of garlic, opuntia, and rubia tinctorum; blue from the ingestion of myosotis palustris, polygonum, mercurials, anchusa, and equisetum. But this azure discoloration must not be confounded with the superficial blue layer which occasionally appears upon milk that has stood for some time, and that gradually permeates the entire fluid if added to milk which is otherwise pure; such milk will not lose its color by triple filtration, as its hue is dependent upon the growth of a fungus (not to be confounded with Hessling's sour-milk fungus), and is identical with penicillium glaucum, and aniline blue; and frequently gives rise to severe attacks of catarrh of the stomach and intestine, and severe prostration. Next to coloring matters, the ethereal oils combine most readily with milk before it leaves the gland. Oil of rape seed, impregnated with sulphur, is quickly absorbed; and in the same way we get the peculiar odor from thyme, wormwood and garlic, when these substances have formed a part of the diet.

The foregoing facts being established, the important question, theoretically as well as practically, arises: How far can disease be propagated through the medium of milk? Not all the chemical and microscopical analyses which have been made for the purpose of solving this question, can lay claim to absolute infallibility. Percy (*N. Y. Med. Journal*, VIII, 1866,) in forty analyses of milk, shows that there is a chemical difference between the milk of well and sick cows, and that important ingredients are entirely wanting in the latter, though he admits he has not been able to demonstrate an active poison therein. A few decades ago, also, was published Hexamer's over-drawn picture refering to the swill-milk scandal in New York, which created a great impression in Europe, though it quickly passed out of notice in New York. Dewees observed yellow fever among nursing women but failed to discover that the disease affected their milk;

D'Outrepont observed the same fact in patients with petechial typhus and I have often made a similar observation in typhoid fever. In diphtheria the chief concern is that infants should not be exposed to direct contagion and that they should approach the mothers only for the purpose of nursing. J. C. Gooding in *Med. Times and Gaz.* 1126, 1872, asserts that unboiled milk, from animals affected with foot and mouth disease, produces derangements of digestion, fever, vesicles and swellings upon the lips and tongue, and marked weakness. There is likewise a published case in which a number of men from an English ship, suffered severely after drinking milk from goats that had fed upon euphorbiaceæ. In regard to the infectiousness of animal flesh when taken into the stomach, authentic reports abound: Gamgee and Livingstone report the flesh of animals which had suffered from epidemic pleuro-pneumonia caused carbuncles and furuncles, and the latter especially emphasizes the fact that boiling and roasting did not destroy this poisonous influence. From this it is evident that goats, sheep, cattle, birds, and fishes may consume many substances which are harmless to them individually, but dangerous to those who afterward partake of their flesh. The milk of a syphilitic mother appeared directly injurious in a case reported by Cerasi (*Gaz. di Roma Jul.*, 1878).

Gallois, Appay, and de Amicis, were unwilling to admit the foregoing, and Banzon's opinion is remarkable, since he is even willing that tuberculous mothers should suckle their young. Bollinger (52nd Vers. d. Naturf. u. Aerzte) on the contrary believes the infection of human beings to be possible through the milk of tuberculous animals. In any event we should avoid using the milk of *old* cows as they are frequently tuberculous, and in all cases the milk should undergo proper sterilizing treatment. Virchow (Berl. Klin. Woch. 1879, 17, 18) does not deny the possibility of infection through tuberculous creatures, and calls to mind the observations made by Kolessnikow (V. Arch. X p. 531) regarding the pathological changes taking place upon the udders of tuberculous cows, which possibly have some influence upon the milk. Uffelmann cites the case of a child that died from tuberculosis, it being impossible to trace the disease to anything but milk from a diseased cow (Arch. f. Kinderh. II), and Stang reports a similar case. Of course it is difficult to point to positive proof in these cases, and more statistics must be collected before we can be absolutely justified in prohibiting the customary supply of milk in every case of constitutional or severe local disease; but the interdiction is warrantable in individual cases, upon the grounds of possibility and probability.

Of importance in connection with the foregoing is the question of transmission of inorganic materials into the milk. While organic chemistry has not yet made sufficient progress to give a decision as to whether a gramme of quinine which gives a bitter taste to the milk, exists in the secretion as quinine or as something else, or whether alcohol, opium, or morphine are again recoverable from the milk, inorganic chemistry, on the other hand, is capable of better results, although here too, there are differences of opinion. The experiments made upon human beings are of a clinical character only. Harnier and Simon found no iron in the residue of human milk, but they readily found salts in loose combination, which quickly disappeared. According to Bistrow's observation, anæmic children improved rapidly, when those who nursed them took iron. According to Wildenstein's experiments, the quantity of milk, under the use of ferric salts, was less, but its specific gravity greater, and the quantity of iron in the ash increased two-fold, but not until it had been in use for twenty-four hours. A small quantity of bismuth was found by Lewald; a large quantity by Chevallier and Henry, and a trace by Marchand. Fifteen grammes of iodide of potassium were found by Lewald in the accumulations of four days; in the following three days twenty-one grammes more were recovered and then all traces disappeared. This was in accordance with a previous calculation. In a further use of iodide of potassium, the milk gave an iodine reaction at the end of four hours, and continued to do so for eleven days.

Lazanski made observations upon a syphilitic mother and her infant five months of age. The mother had been infected two years previously, had no indications of the disease upon her genital organs, but had syphilides in the groins, and swollen glands; the child was affected upon the skin and the mucous membranes. The mother received half a gramme of iodide of potassium twice a day, the result being that the urine gave an iodine reaction upon the same day that the treatment was begun, and in the child the iodine reaction appeared upon the following day. Gemmel also relates that a cow that received ten grammes of iodide of potassium daily, began to eliminate it through the milk glands on the tenth day. In a case where the nurse was treated with iodide of potassium, a desired effect was soon produced upon the child which she was suckling. Upon the basis of such facts, Leviseur recommends in the secondary syphilis of infants, the use of iodide of potassium through the medium of the breast-milk, likewise the sulphate of quinine in neuroses of an intermittent character, and arsenic for the moist eruptions upon the skin. Arsenic was found seventeen hours after it had been given, and it continued to be traceable for sixty hours thereafter. Hertwig states that medicinal doses for a cow are sufficient to poison the meat. Lead may be found in milk, likewise oxide of zinc, and probably all other preparations of the same metal; they were found in from four to eight hours after they had been given, and disappeared after fifty or sixty hours. Antimony passes into the milk, therefore special care should be exercised in prescribing it. Mercury has not been found by Peligot, Chevallier and Henry, and Harnier, but it has been traced by Lewald and Personne.

O. Kahler made the cases of three women who were receiving mercury by inunction, the occasion of accurate investigations. The milk was examined by the chemico-electrolytic method of Schneider, but no mercury was found; he considers the affirmation of Lewald and Personne, under this head, as questionable. In my personal experience, the results of mercurial treatment of the mothers and nurses of syphilitic infants, where the disease was hereditary, has not been satisfactory; but in cases where the symptoms of the disease first appeared after the child was some months old, the customary internal treatment yielded very beautiful results. Tudakowski was able to detect traces of mercury in three hundred and sixty-six grammes of milk tested according to Schneider's method. Likewise, Klink treated a syphilitic mother with twenty-five inunctions of ung. hydrarg., giving twenty grammes at each inunction. Her infant had large condylomata and adenitis, and the latter quickly improved (during the same period the infant had three baths, each containing 0.3 gramme of corrosive sublimate). Carbolic acid, bicarbonate of potassium, chloride of sodium, Glauber's salts and sulphate of magnesium are all transmitted in the milk. The vegetable acid salts develop carbonic acid in the milk, but the alkaline compounds of sulphur, according to Marchand, do not. Stumpf found iodine speedily in the milk of women, but slowly in herbivorous animals; it is found in combination with casein, but in uncertain quantities. Alcohol he did not find in the milk of herbivores, but lead was discovered in traces, and salicylic acid in small quantities. A large number of similar observations and experiments have since been made, too many to be here recorded; but one of the most interesting is that of Dr. Koplik, who observed iodine eruptions in a baby, whose mother took iodide of potassium in but small doses.

DIET OF WET-NURSES.

If, after all that has been said we have succeeded in fairly bringing forward the question as to how the nourishment of a nursing woman should be regulated in order to obtain the most desirable ends, the answer thereto can be given in all its details. Powerful salts are to be avoided under all circumstances, as well the salines and more powerful drastics; likewise an injudicious use of table salt, ethereal oils, and strong condiments. Furthermore everything is to be avoided, which has a tendency to derange or to weaken digestion and assimilation. Generally the wet-nurse looks upon her position as one "which flows with milk and honey," where "roasted pigeons fly into the mouth," and no end of good things to eat until the appetite is satisfied—or spoiled. Somewhat more of albuminous food is indicated than under ordinary circumstances, though too much of this, or an exclusive diet destroys health, and the milk secreting power as well. Much fluid food, and an abundance of water, will increase the volume of the milk; a moderate use of beer may act as a stomachic; water and barley-gruel act upon the milk by virtue of their fluidity as well as nutritious qualities, and the same is true of tea in moderate quantities. Potatoes in large quantity, and other carbo-hydrates, are to be avoided as a principal means of nourishment, but fat in moderate amount is desirable. In general it may be laid down as a fundamental principle, that a wet-nurse will have the largest quantity, and the *best*, milk, when using the same nutriment to which she was accustomed before pregnancy (provided it keep her in good physical condition), with the addition of a certain quantity of albuminoids, and plenty of fluid food. What the nursing woman expends as milk, must be restored to her. The daily requirements of a woman who is not doing hard work, nor suckling a child, are: Albumin, gm. 85.00; fat, 30.00; carbo-hydrates, 300.00. The requirements of a nursing child of five months who receives a daily allowance of 800 grammes of breast-milk are: Albumin, gm. 20.00; fat, 31.00; sugar, 48.00. This quantity must be obtained from the reservoir of the mother, and is best obtained from a richly albuminous diet. Where we are dealing with small atrophied breasts, and it is necessary to stimulate the secreting function, considerable time is required before a free and satisfactory flow of milk can be obtained. If we are looking simply to keeping up an abundant supply, carbo-hydrates may prove satisfactory. The most common question is, however, in regard to the improvement of the substance of the mammary glands themselves. It must not be forgotten that tissue changes and good health do not depend alone upon what is eaten. A wet-nurse must not be thrown too suddenly upon conditions which are quite strange to her. She must live, as nearly as possible, in the manner to which she has been accustomed. A nurse who is removed from the hay-field or kitchen to the boudoir, and who is held in restraint from fear lest she might eat a raw apple, drink a glass of beer, meet her lover, or who is deprived of her customary physical exercise, will not retain her health nor give a proper quantity of milk. It is in accordance with these fundamental directions, that the various articles must be criticised which have been recommended as proper means of nourishment and diet during the period of lactation. The list of such articles contains beer, buttermilk, milk, chocolate, thick soups, husked grains, oysters, crabs, sea-eel soup, etc. If all these dietetic means do not accomplish their end, one has to look around for therapeutic measures for the stimulation of the milk-secreting function, with more or less of reason, and more or less of confidence. C. Gesner, ("Concerning Things which Make the Milk Plentiful "* p. 45), has referred to all the materials used, from Galen's down to his own time. Hufeland recommended a milk-

---

* "De his quæ lactis ubertatem faciunt," 1546.

making powder, Moleschott the edible chestnut, Routh and Gillian the leaves of ricinus communis, and in England and America, the latter applied to the breasts have won a popularity far exceeding their value. The list of galactagogues which Routh published, makes a very respectable appearance, but it cannot be said that they can boast of great success. I have used the electric current many times to stimulate the flow of milk, and with good results, but the galvanic current, with its influence on circulation, has a better effect.

SUBSTITUTES FOR BREAST-MILK.

Infants deprived of breast-milk, that have never had it, or secure only an insufficient supply, require artificial feeding. The substitutes should be as near normal woman's milk as possible; and naturally, when the latter cannot be had, animal milks are selected. Of these, only two are practically available—those of the cow and the goat. The objection to either is chiefly valid in regard to the latter, for the chemical incongruities and other difficulties, to which I shall allude in regard to cow's milk, are even more pronounced in that of goats.

MILK OF ONE COW.

The milk of the cow is not necessarily uniform and unchangeable. Its nitrogenous constituents vary; its composition, (and at all events its taste), is frequently altered by changes in feeding, or by the accidental admixture of odoriferous or purgative herbs. The nursling takes the whole contents of the mother's breast; from the udder of the cow it gets but a small portion, which varies for different reasons. The first period yields milk containing less fat than that which is obtained towards the end of the milking; besides, milk taken from a pail, contains more fat in its upper part than in the lower. Pasture, or dry feeding of the same cow, induce differences in her milk; thus the dairies established to supply infants and children with cow's milk, in Frankfort and other places, require uniform stable feeding throughout the whole year. Diseases of the cow influence the milk considerably; tuberculosis is frequent. Thus a baby fed on the milk of one cow is, as it were, an appendage of, and dependant upon that animal; consequently the milk of one cow is inferior to that of a whole dairy, for by the latter we dilute and diminish dangers which attend changes through feeding and sickness.

RAW OR BOILED MILK.

Boiling removes a small portion of fat and casein which is collected on the surface, a loss that is desirable, though the quantity withdrawn is usually too small. Still the removal of fat by allowing the milk to stand is improper, inasmuch as during the time necessary for that purpose the milk will acidulate. Boiling retards acidulation, and neutral milk becomes alkaline through boiling; further the formation of lactic acid is delayed through the expulsion of a large quantity of gases contained in the milk when it leaves the udder.* True, boiled milk is less pleasant to most, but its possible dangers are less to all; still, with the "volatile principle which is destroyed by boiling, of unknown nature, but presumably beneficial effect," which is good-naturedly talked about by some authors, I have but little sympathy. Milk changes very readily; it takes the odor of substances near by, it is more than merely suspected of communicating contagious diseases by carrying bacteric poisons, and it acidulates very rapidly. Bechamp has observed the formation of alcohol and acetic acid within the udder, and Hessling reports milk so impregnated with fungi that coffee to which it was added became poisonous. Most of these changes are retarded, interrupted, or annihilated by boiling, and any apparatus is good enough for the purpose. After boiling, milk destined for the use of a baby during the day should be kept in clean bottles containing from three to six ounces, filled up to the cork, and the bottles then turned upside down in a cold place; such will keep longer than milk preserved in the usual way. Before using, it may be heated in a waterbath; and by repeating this heating of the whole amount of the day's milk, several times

---

* Three per cent., according to Hoppe (carb. acid 55, 15, Nitrogen 40, 56, oxygen 4, 29). According to Pflueger's arch. II, p. 166, oxygen c. 1–0.09 per cent., nitrogen 0.7–0.8 per cent., carbonic acid 7.6–7.4 per cent.).

during the twenty-four hours, fermentation will be retarded, and digestibility improved.

CONDENSED MILK.

There are condensed milks in the market which claim to contain no sugar ; others manufactured for the purpose of keeping an indefinite time, which have a percentage of nearly fifty per cent. The milk distributed in New York for immediate use, contains from eleven to thirteen per cent. Some manufacturers condense pure milk, others find it more in accordance with their bank accounts to use skimmed milk; thus, Soxhlet found but 60–80 parts of fat (instead of 100–110), for every 100 parts of nitrogenous substance. Now, what is "condensed milk"? Can it be expected that the great mass of the public will always be careful in selecting the same brand, though it is possible that the product be always uniformly prepared?

The sugar contained in the condensed milks is mostly cane sugar ; in some condensed "Swiss" milks 14-18 per cent. of milk sugar and 24-30 per cent. of cane sugar (Werner and Kofler). Thus we cannot always be sure of the nature of the sugar in the article, much less of its quantity. Though Nature allows of latitude, that required by the several manufacturers is greater than is convenient or admissible. At all events, the different composition of condensed milks explains the variety of opinions expressed of their effects. It has been stated that the formation of lactic acid is increased by it (Kehrer); that it gives rise to thrush and diarrhœa (Fleischmann); that it fattens children but predisposes them to rhachitis (Daly). I fully agree with those authors who discard condensed milk in all cases where cow's milk is to be had; but in large cities, the choice is frequently between two evils—bad milk and condensed milk. In such a case I permit condensed milk. But it should not be used in the diet of babes when fair and unadulterated cow's milk can be had, and this is possible, if the same pains and some little of the money which is now used for condensing, were spent on supplying the children of the poor, by rich philanthropists.

OTHER MILKS.

The frequency of ill success when cow's milk is fed has given rise to many experiments with the lacteal secretion of other animals. Koenig has the following analyses:

	Water.	Casein.	Albumin.	Total nitrogenous subs'ances.	Fat.	Milk-sugar.	Salts.	No. of Analyses.
Sheep	81.63	4 09	1 42	6.95	5.83	4.86	0.71	15
Lama	86 55	3.00	0.90	3.99	3.15	5.60	0.80	3
Camel	86 94	....	....	3.84	2 90	5.66	0.66	2
Mare	90.71	1.24	0.75	2.05	1.17	5 70	0.37	27
Ass	90.04	0.60	1.55	2.01	1.39	6.25	0.31	17
Sow	84.04	....	....	7.23	4.55	3.13	1.05	9
Bitch	75.44	5.53	4.38	11.07	9.57	3.19	0.73	16
Cat	81.63	3.12	5.96	9.08	3.33	4.91	0.58	1

According to Jacquemier, there are in one thousand parts of woman's milk 26.66 fat, 39.24 casein, etc., and 1.38 salts; in bitch's milk 97.20 fat, 117 casein, etc., and 13.50 salts. The observations on the effect of the latter are but few in number, and appear to have been recorded with a certain degree of preoccupation; thus young dogs fed on woman's milk are said to have died of diarrhœa, while rhachitic children fed on bitch's milk are reported to have recovered in an incredible time, even to the straightening of curved limbs.

Mare's milk is very much like asses' milk; its reaction is alkaline and remains so for days. When it acidulates, the casein is thrown out in small delicate flakes soluble in acids. The casein thrown out by alcohol is finely flocculent, like woman's casein; even when dried and deprived of fat, the flakes remain yellowish and loose. (Cow's casein becomes hard.) Mare's casein is less soluble in water than woman's, but more so than that of the cow. In artificial gastric juice, mare's and woman's casein behave equally. Thus in cases of necessity, or opportunity, mare's and asses' milk would be proper substitutes for that of woman.

Goat's milk contains more fat, casein, albumen and salts, and less sugar than either cow's or woman's milk; its drawbacks as nutriment for the infant are therefore more serious than those of the latter, besides, it has an unpleasant odor. I never saw infants taking it for any length of time, and seldom knew it to agree with them nearly so well as cow's milk.

The goat has been enthusiastically eulogized by Boudard; he is particularly stricken with the white, hornless goat for its "odorless milk, sweetness of temper (*douceur de ses moeurs*), love of the protecting stable life, aversion to freedom which might expose her to enemies, large round eyes, sentimental mien, and classically formed nipples." It permits the babies to be put to the nipple, it has been recommended by Buffon and Saint Vincent de Paul, and its "heraldic emblems have been immortalized by heaven, air, earth and oceans." Poor babies, to be fed on rhapsodies!

The main objection to goat's (and cow's) milk is to be found in the *large percentages of casein and fat.* Of casein, according to Biedert (and general experience) an average of *one per cent.* is all that can be digested. Of fat, every milk—even woman's—contains more than appears to be required for physiological purposes. This is the reason why Ballot recommended buttermilk as infant food, which is composed of water 90.62, nitrogenous substance 3.78, fat 1.25, milk-sugar 3.38, lactic acid 0.32, and salts 0.5. In it, evidently, fat is reduced in quantity, but casein, unfortunately, is not, and the additional lactic acid is a detrimental feature. Of buttermilk, Ballot cooks a quart with a tablespoonful of wheat flour (rice flour in cases of diarrhœa) for some minutes, and adds a gramme (grs. xv) of sugar. With this he begins the third week after birth, and succeeds in proving that many liberties may sometimes be taken with human digestion, without apparent immediate injury.

FAT.

Fat, as contained in milk, is by no means a simple substance, but consists of at least nine compounds; it is probably identical in the different varieties of milk. It has been studied most extensively in the secretion of the cow, where it is found in different percentages, There is more of it in colostrum, in the upper layers of the milk contained in a pail or bowl, and in the evening. Its average percentage is 4.3, which is a great deal more than is contained in woman's milk. Thus I have always taught that it is better to reduce the fat of the milk which is to be given to infants than to add to it. Still it is claimed that this figure is incorrect, particularly by Biedert and A. V Meigs.

The methods of reducing the amount of fat in cow's milk differ. To allow it to rise spontaneously, so that it can be *skimmed* off, allows lactic and fat acids to form, even in low temperature; and skimmed milk has a specific gravity of 1035-1037, and contains water 90.63, nitrogenous substances 3.06, fat 0.97, milk-sugar 4.77, and salts 0.75. Thus *skimmed milk prepared or obtained in the usual way is objectionable* as food for the young; but milk deprived in part of its fat by centrifugal machines which terminate their work in half an hour, may be employed to advantage. Normal milk, containing water 87.79, casein 2.73, albumen 0.68, fat 3.64, sugar 4.69, exhibited after the process of centrifugal separation, water 90.73, casein 2.88, albumen 0.49, fat 0.46, sugar 5.34, and salts 0.72.

CASEIN.

The high percentage of casein is not the only objection to the substitution of diluted cow's, for woman's milk. There are important differences in the chemical and physiological differences of the two caseins.* From the physician's point of view it is important to notice that solutions of mineral acids—lactic, acetic and tartaric—sulphate of magnesia, and phosphate of calcium, do not coagulate woman's as they do cow's milk. Coagula of woman's milk in the stomach will dissolve in a surplus of gastric juice; cow's milk will not behave so, according to Biedert (contrary to Putnam's observations, who denies the existence of this difference). Thus cow's casein is not nearly so digestible as woman's milk, and, as before stated, there ought to be but one per cent. of such casein in an infant's food. To thus reduce the quantity of casein, and restore the equilibrium between casein and fat, Biedert has prepared a cream mixture, which, in a condensed form, has found its way into the market. His schedule for its preparation is changeable for every month of the infant's life, and reads as follows:

---

* A. Jacobi, in Gerhardt's Handb. d. Kinderk., 2, Ed. I 2, p. 91. Bunge in Jahrb. f. Kinderh. 1875, IX 1.

	CREAM.	MILK.	WATER.	MILK-SUGAR.	CASEIN.	FAT.	SUGAR.
1st month	℥iv.		℥xii.	℥ss.	1 p.c.	2.4 p.c.	3.8 p.c.
2d month		℥ij.			1.4 p.c.	2.6 p.c.	3.8 p.c.
3d month		℥iv.			1.3 p.c.	2.7 p.c.	3.8 p.c.
4th month		℥viii.			2.3 p.c	2.9 p c.	3.8 p c.
5th month		℥xii.			2.6 p.c.	3.0 p.c.	3.7 p.c.
6th month		℥xvi.	℥viii.	ʒiiss.	3.2 p.c.	3.0 p.c.	4.0 p.c.

These mixtures are ingenious modifications of Ritter's formula first proposed in 1863, which adds two or more parts of water to one of sweet cream, or Kehrer's, who uses one part of cream with two of whey.

A. V. Meig's cream food is prepared in the manner described in his book on "Milk Analysis and Infant Feeding," 1885, p. 74–76. I copy literally, because the deserving author ought to be heard in his own words. Still I take the liberty of marking such words or sentences as I want my readers to consider in regard to the difficulties of preparing the food, the time it takes, the intelligent management it requires, the expense it occasions, compared with what I claim are the absolute requirements for every infant food, viz., *accessibility to the poorest in pocket and brains* in the land, *simplicity, cheapness and digestibility.*

"There must be *obtained from a reliable druggist* packages of pure *milk-sugar* containing seventeen and three-quarter (17¾) drachms each. The contents of one package is to be dissolved in a pint of water, and it is best to have a bottle which will contain just one pint, as there is then no need for further measuring.

The contents of one of the sugar packages is put into the bottle, and when filled with water soon dissolves, and is ready for use. The dry sugar keeps indefinitely,· but once dissolved *it sours if kept more than a day or two, in warm weather;* it is understood, therefore, that the sugar-water must be *kept in a cool place*, and if it should at any time become sour (which *is easily discovered by smell and taste*), *should be thrown out*, and after the bottle has been carefully washed with boiling water, the contents of a fresh package dissolved. A milkman *must be found who will serve good milk and cream, fresh every day*. By good milk is meant ordinary milk, such as is *easily procured in most cities*, and not rich Jersey milk; and in the same way the cream should be such as is *ordinarily used* in tea and coffee, and *not the very rich cream* of fancy cattle. The reason that ordinary milk and cream are recommended is *because they are within the reach of almost every one*, and not because they are any better than the rich milk of high-priced stock. If Jersey milk was to be used, it would be *necessary to analyze specimens, and then make the necessary calculation*, as to how to dilute it to obtain the desired relative proportions of the proximate principles. When the child is fed, the nurse should mix together two (2) tablespoonfuls of cream, one (1) of milk, two (2) of lime-water, and three (3) of the sugar-water, and then, as soon as the mixture has been warmed, it may be poured into the bottle, and the food is ready for use.

If the infant is healthy, this quantity will not satisfy it after the first few weeks, and then double the quantity must be prepared for each feeding. Twice as many tablespoonsful of each of the ingredients must be mixed together, making sixteen tablespoonsful (about half a pint) in all."

In a paper "On the Artificial Feeding of Infants," read by the same author before the American Pediatric Society assembled in Washington, September 21st, 1889, he says: "Instead of taking cream and milk in the proportions respectively of two and one in eight, three parts of a weak cream may be used, which is obtained as follows: one quart of good ordinary milk is placed in a high pitcher, or other vessel, and allowed to stand in a cool place for three hours; then one pint is slowly poured off from this, care being taken that the vessel is not agitated, the object being to obtain the upper layer of fluid, rich in fat, and leave the lower, comparatively poor portion, behind. When the child is to be fed, there are taken of this weak cream three tablespoonfuls, of lime water two tablespoonfuls, and of sugar water three tablespoonfuls. The sugar water is to be made in the proportion of eighteen drachms of milk sugar to one pint of water. This makes only four ounces of food, and if the infant is old enough to require eight ounces at once, double the quantities of each of the ingredients must be mixed. This is simply warmed in a bottle as usual, and is then ready for use."

Biedert himself does not claim that his cream mixture is indispensable; on the contrary, he says in the very circular which accompanies the preparation, "The first requirement of an infant food is, to prove digestible for all infants, including the feeble and sick. Such a food must not contain more than one per-cent. of casein, therefore milk must be mixed, for the very young, with three or four parts of an indifferent food; this latter quantity must be gradually diminished. This mixture is far superior to any mixtures and artificial preparations thus far prepared, particularly those which contain large amounts of flour or sugar."

As the percentage of fat changes with that of casein in this mixture, Biedert concluded that it is necessary to add fat in those cases, in which the other was not well tolerated either by the well or the sick. for the latter he particularly recommends until recovery renders the return to the original mode of feeding advisable. This temporary administration cannot prove very expensive. As for the fact that the cream mixture is more expensive than the simple food recommended as a rule, he alludes to the expensiveness of a wet-nurse for which the cream mixture may be substituted from birth.

My readers will notice the modesty and unobtrusiveness in Biedert's expressions upon, and recommendations of, his preparations, particularly of the condensed mixture which is offered for sale. In this lies the difference between a manufacturer who, in laboring for pecuniary gain by working on the credulity of the easily deceived public, announces his wares as panaceas for every ill, and cannot see how the world went on before a benign fate sent him to redeem it, and the humane physician of scientific instincts who recommends his gifts for what he believes or knows them to be worth. It is self-understood that the author himself has no monetary interest in the sale of his "Cream Preserve" (Rahm Conserve). The mixture of Biedert, indeed every cream mixture, requires that its emulsion should be complete; in one in which it was not so, the fæces exhibited fat to the amount of forty per cent. Such occurrences must prevent the introduction into common use of such fat emulsions, for women at large, will not spend the required time and pains in their preparation; besides, cream is not the same thing always, and if accuracy in measuring percentages while nursing be really as indispensable as claimed, the irregularity in composition is a serious obstacle to its use. On account of these very variations Biedert was induced to compound his "Cream Preserve," which keeps for some time. When used after three or four weeks, some milk has to be added.

An excess of casein in milk is justly feared for its irritating influence. An excess of fat works in the same way. Even normal mother's milk leaves a high percentage of fat in the infant's fæces as normal movements contains twelve per cent. of fat. This percentage, when Biedert's mixture was administered, varied from 20.3 to 3.8, and there is no report that this low percentage was complicated with anything like bad symptoms. The higher percentage, however, is uncomfortable and dangerous. Arthur Meigs (Milk Analysis and Infant Feeding, Philad. 1885, p. 63) quotes a similar remark of mine and adds: "It seems scarcely wise, upon such theoretical grounds to condemn the use of cream, particularly when the experience of many physicians has been that they were able to feed successfully upon cream, or cream mixtures, children they were unable to manage in any other way. Surely, Dr. Jacobi would not condemn the use of human milk, because fat may be found in the stools !"

Certainly not. It is not that it *may* be found, but it *is* found. That is no theoretical reason, but a universal fact; it is a better established fact than the experience of *many* physicians who were able to feed upon cream. What I do say is, that as the infant intestine does not digest by far the quantity of fat contained in the normal infant food, we ought not to increase the amount of the indigestible (and in their high percentage, useless) ingesta by deliberately adding to them. Biedert, himself a careful experimenter and the persistent advocate of his cream mixture, points out the danger of giving too much fat. It is he who added the chapter of "fat diarrhœa" to our pathology, and who advises the diminution of the percentage of fat in the food for the purpose of relieving that morbid condition.

ADDITION OF WATER TO COW'S MILK.

The dilution of cow's milk, boiled or unboiled with water, with or without the addition of sugar, was naturally the first attempt to make cow's milk similar to that of the human female. The thousands of recommendations as to definite proportions and percentages, distributed as they are through books and journals, are only repetitions of what all mothers are accustomed to do. In many large lying-in institutions nothing is furnished for the nutrition of the infants but milk and water; this is the way in which they were fed, according to Parrot, at the Hospices des Enfants assistés, and Pfeiffer says, " The nursing child ought to have milk, and milk alone."

Natalis Guillot (who was the first to introduce a method of systematic weighing for the purpose of ascertaining the influence of nourishment) placed his figures entirely too high, in demanding 20–25 daily nursings of 25 grammes each time; but Bouchard came to the more correct estimate of 8–10 nursings, of 3 grammes each, of mother's milk, upon the first day of life, of 15 grammes each upon the second, of 40 upon the third, and of 55 upon the fourth; from the fourth day the quantity to be increased slowly.

According to the same author, the quantities may be tabulated as follows:

1st day.................... 30 grammes (1 ounce).
2d day.................... 180 grammes (6 ounces).
3d day.................... 450 grammes (15 ounces).
4th day................... 550 grammes (18 ounces).

After the first month 650 grammes daily, after the fourth 850, between the sixth and the ninth 950. These figures Jacquemier has taken as the foundation of his directions for the dilution of cow's milk. He mixes two parts of milk with one of water, this being the same as used by Parrot in the Hospice des Enfants assistés. In this way the child gets 20, 100, 300 and 366 grammes of milk upon the first, second, third and fourth days respectively. After the first month the quantity is 434 grammes daily, after the third 460, after the fourth 566, between the sixth and the ninth 634, in all of which cases half as much water is added. This dilution has been approved of by many authors and practitioners, with the limitation that in the case of very small children the quantity of water is increased; for those who are larger it is diminished.

Now comes a question as to the influence of an excess of water. Is it harmful, unimportant, or useful? Is water useless ballast, annoying the skin and kidneys, and increasing the number of soiled diapers, or has it another significance?

Water and urea stand in fixed relations to each other. Bischoff (Der Harnstoff als Maasstab des Stoffwechsels, 1853) found urea increased in man and in dogs in proportion to the quantity of water passed through the kidneys. Genth arrived at similar results in experiments made upon himself (Unters. üb d. Einfluss d. Kochsalzes, u. s. w., 1860), but also ascertained that the quantity of urea is increased only when the quantity of water passed through the kidneys is increased (Zeitsch. f. Biol., 1866, p. 338). E. Wolff says "we must strenuously avoid in practice all causes which can promote greatly increased consumption of water, such as watery food, too much salt, high temperature, as in this way many of the means of fattening are lost" (Fütterung der Hausthiere, 1876, p. 310). But neither child nor man is fed simply for the purpose of being fattened like animals. Water stimulates physiological changes. The excretion of sulphate of potassium, or of sulphates in general, of phosphate of sodium, and phosphatic earths, of chlorides and of urea is increased, likewise the discharge of excrementitious matters from the intestines, and of carbonic acid. In proportion as this takes place, the volume of the solid substance of the blood increases. The serum loses water and chloride of sodium, and tissue changes become less rapid. With a fluid diet the temperature is lowered, the heart pulsations and the action of the lungs are retarded. When no water is taken, oxalic acid accumulates in the blood. Of the water which is introduced into the blood, a portion enters the blood globules. The greater the percentage of solid constituents the greater is the proportion of water which is given off to these solids, and the more chloride of sodium and other salts will be released by the blood-cells. Of course these products will bear no absolute relation to the quantity of water taken up. The addition of fifty times its volume of water produces a variation in the quantity of water absorbed by the solid constituents of from 12.2 to 19.8 per cent. But these changes are great enough to influence favorably the tissue changes (H. Nasse, Arch. f. d. ges. Phys. xvi., 1878, s. 613).

The free dilution of children's nourishment with water is demanded upon the following additional facts. Only to a certain limit will pepsin be furnished for digestive purposes, and probably a portion of this is not entirely utilized. A great quantity of water is necessary to pepsin digestion. In artificial digestion albumin often remains unchanged until large quantities of acidulated water are supplied. Without doubt many disturbances of digestion are to be explained by a deficiency of water, certainly many more than are due to an excess of it, as it is so quickly absorbed.

For the reasons given, I advocate under all conditions a plentiful addition of water to children's food, and in this connection I lay stress upon the fact that, as a rule, small children receive water only as they get it in their milk. Alike in summer and in winter, it is probable that the fact seldom occurs to a mother or nurse that a child can be thirsty without being hungry at the same time. Certainly many a discomfort and even illness is conditioned upon the fact that children have been compelled to eat in order to satisfy thirst, and often to suffer thirst because over-stimulated and injured stomachs will take no more nourishment at irregular and too short intervals. There are also normal products of digestion capable of producing disturbances in the digestive process, chief among which is peptone itself. I have, therefore, considered it necessary in preparing rules for the feeding of children, which the New York Board of Health annually publishes, to insist upon giving infants an occasional drink of water, at least during hot weather. When there is the least ground for the supposition that the drinking water is contaminated with germs of disease, or where it is unusually hard, it should be boiled before its admixture with food, whether the diet be of milk or a mixed one. In general for very young infants, it will give greater satisfaction to use the boiled water systematically even where there is no apparent urgency for it.

ADMIXTURE OF SALT.

The rôle of salt in the matter of nourishment is an exceedingly important one. It is generally known that animals are made greedier for a proper quantity of food by the addition of a moderate quantity of salt, (say 30—50 grammes 1—1½ ounces daily for each thousand pounds of live weight) and that the food thereby gains not only in palatableness, but also apparently in nutritiousness. " On the other hand, we know with little definiteness whether salt increases the digestibility of the food as a whole, or whether it has no such effect." (E. Wolff.)

It is quite certain that by the active tissue change which is accomplished by salt, desire for food is stimulated, and that which is already in the intestines is better retained and absorbed, and the residue is discharged more thoroughly altered. Carnivoræ do not need salt; herbivoræ do need it because, although the other mineral elements necessary for the animal organism are provided in their food in sufficient quantity, and in a more or less proper condition of preparation, the required amount of salt is wanting. Particularly is the relation of sodium to potassium disturbed. In the food of carnivoræ and of herbivoræ, the *absolute* quantity of sodium and of chlorine is very nearly the same, but in the food of herbivoræ there is two to four times as much potassium relatively as in that of the carnivoræ. Therefore, there is found in the blood of herbivoræ an abundance of potassium salts (posphates, etc.), which needs antagonizing by sodium for the purpose of elimination.

When Bunge took large quantities (18 - 24 grammes) of phosphate and citrate of potassium for four successive days, he lost half of all the sodium in his circulation. On the following day very little was discharged, because he had but little left. (*Zeitsch f. Biol.*, IX, 104, 1873; X, 227, 295, 1874.)

In the body of the child the physiological effect of salt is all-important, no matter whether it is directly introduced through the mother's milk, or added as a condiment to cow's milk and vegetable diet. A portion of that which is introduced may be absorbed in solution; another portion, however, will certainly be broken up into another sodium salt, and into hydrochloric acid. (Beneke.) This normal ingredient of the liquid contents of the stomach, serves as an excitant to the secretions of the glands, facilitates digestion, and stimulates appetite. The excess of acid which may get into the intestinal canal, unites with the sodium of the bile in the duodenum, assisting in producing a second combination of chloride of sodium, and this is again dissolved in the intestines. From this point another active usefulness begins; it consists of osmosis from the surface of the intestine to the intestinal villi and blood-vessels; from the villi it passes into the substance of the blood, from the serum into the corpuscles; from the blood it goes into the tissues, and out of them, to final elimination. It is chiefly chloride of sodium, other solutions not excepted however, which renders the generation and disintegration of tissues possible. The effect of moderate doses of salt is evident both to the chemist and to the clinician. It is also observed that, without any more water being taken, the kidneys will have more powerful action, and the quantity of urea will be increased, which means that the metamorphosis of albumin is increased (about 4.7 in ordinary conditions) in consequence of the increased circulation of the parenchymatous liquor. Of course this circulation bears a relation to the quantity of salt absorbed; large quantities hasten the changes in the albuminates and thereby necessitate additional water; in that way they again increase the urea and carbonic acid discharges; besides, they excite the surface secretions to an unusual degree.—They act not merely as diuretics, but also as laxatives. In the *Jour. f. Kinderkr.* for 1873 there is an account of a specimen of mother's milk which contained eight per cent. (! !) of salt, thereby jeopardizing the life of the infant through violent diarrhœa ere the cause of suffering was discovered and removed.

I return to the fact that vegetable diet contains more potassium and less sodium than milk, and the milk of herbivora more potassium than that of the carnivora. Cat's milk contains sodium 1. to potassium .76, woman's milk 1. 1.13 to 4.4, sheep's and cow's milk 1 to 5.6. The quantity of salt in woman's milk is influenced greatly by the direct addition of the same to the food. These facts are of great importance in the preparation of an artificial diet, whether of vegetables or of animal milk, designed for the human infant. The addition of salt is not only of great physiological importance in the interest of tissue changes in general, but without such addition artificial diet is deprived of the chemical mixture from the beginning, which renders it quite similar in this respect to natural diet. An extremely important fact also is, that the addition of chloride of sodium delays and renders difficult the firm curding of milk by rennet (Pflüger's Arch., XIII, p. 93); thus it should be added to cow's milk as a general rule, and to woman's milk when the large curds brought up by vomiting, or evacuated by rectum, exhibit an undue amount of coagulation.

ADDITION OF SUGAR TO COW'S MILK.

The quantity of sugar in the milk of the woman, the ass, and the mare, is greater than that in cow's milk. Immediately after the milking, perhaps even before it, lactic acid change begins, a process which, together with the gradual conversion of fat into acids, is the cause of the curding. Its high percentage in woman's milk, with a smaller percentage of casein and butter, gives a blueish color, and to the colostrum (which, besides, is rich in salt) its tendency to produce diarrhœa. Sugar occasionally manifests itself under abnormal conditions. In the milk of anæmic women it is occasionally found to an unusual degree, in which case other solid matters may be diminished, though this does not always follow; under such circumstances the infant not unfrequently exhibits obstinate diarrhœa.

The conversion of milk-sugar into lactic acid takes place very rapidly, by which cow's milk is made sour at once. Cane-sugar is not so easily transformed, and is sometimes utilized for the purpose of counteracting the rapid conversion of milk-sugar, and the preservation of articles of food in general.

Condensed milk remains intact a long time on account of the addition of cane-sugar in spite of the presence of milk-sugar. Therefore, it is not at all an indifferent matter, *whether milk-sugar or cane-sugar is added to the artificial diet.* Still the use of the milk-sugar has been urged upon other grounds. It has been stated that milk sugar is to be preferred because it contains phosphatic salts; but verily these may be introduced into the body in other and less dangerous ways; phosphates are so plentiful in all foods, that they are eliminated as fast as introduced. It has also been said that milk-sugar is contained in milk by virtue of a natural arrangement, and if Nature had intended to have cane-sugar in milk, she would have supplied cane-sugar. Nevertheless it is a fact, often a disagreeable one, that milk-sugar is quickly converted into lactic acid, so that an excess of acid accumulates in the stomach, and causes the protein substances to coagulate and become indigestible; it dissolves the alkalies and the calcium out of the phosphate combinations to no purpose except to produce dyspepsia and diarrhœa, and (according to Heitzmann and others) rhachitis. These facts furnish reason enough for carefully *avoiding the use of milk-sugar, and for preferring cane-sugar as an addition to cow's milk* and artificial foods.

Above all, it has long been known that cane-sugar is partially converted into grape sugar while yet in the stomach, since Uffelmann observed in the stomach of a boy upon whom gastrostomy had been performed, an abundant conversion of cane-sugar into grape-sugar beginning within forty-five minutes. In accordance with what has been said, I therefore assert that artificial nutriment for children should be mixed with *cane-sugar*, not with *milk-sugar*, in view of the active production of acids in the young, their tendency to diarrhœa, and the danger incurred by the premature discharge of salts.

ANIMAL ADMIXTURES.

Human knowledge and foresight have almost exhausted themselves in efforts to procure a reliable cow's milk. The result of all is *good cow's milk, but no equivalent for mother's milk*. But I do not mean to say that cow's milk, or cow's milk mixed with water, is absolutely poisonous and injurious to babes; far from it. The fact that there are infants who thrive on that exclusive diet, and whose general condition remains such as not to require medical interferences in their behalf, would give the lie to such an exaggerated opinion. Still, there are a great many infants who *appear* to thrive better than they *actually do*. Many grow fat and rotund without apparent harm on improper food, still we find many who, while increasing in weight, gradually lay the foundation for future ailing; for such fatness and rotundity means, too frequently, rhachitis, and requires watching and change of food. The number of infants, however, remaining in good condition on that food is not very small, hence what does it prove? Nothing, but that digestive processes permit of a certain latitude, that Nature does no routine work, and that the sum total of vital processes does not respond to certain influences like reagents in a chemical test, where the same procedure always yields the same results. There is *no food*, unless quite absurd, on *which certain infants will not thrive*. But when, in a large percentage of cases, the unfortunate results are established, and when these can be traced to their exact and uniform cause, that particular food is to be condemned. The assertion that cow's milk is an exact substitute for human milk, is the counterpart of that which obtained equality for human and animal blood. The practice, based upon this assumption, and resulting as it did in the transfusion of animal blood into the human circulation, out-lived itself very speedily.

Such practitioners as convinced themselves of the ill-success often attending the use of milk, or watered milk, commenced at an early period to mix with meat-soups, meat-tea, or egg. Bretonneau reported as early as 1818 that "tabes mesenterica" disappeared in the hospital of Tours from amongst children fed on beef soup and milk; and this mixture Vauquelin declared to come nearest to mother's milk of all preparations. The administration of some beef soup, well made, a cupfull every day (mutton broth when there is a tendency to diarrhœa), is advisable towards the end of the first year. Long before this period, indeed at any time during infancy, it is indicated in cases of early rhachitis, rhachitic constipation, undue adiposity, and retarded teething.

Beef-essence, well made, in a bottle swimming in a waterbath, is still deemed by some to be the model food. That it is not so rich in soluble albuminoids as was believed, ought to be generally understood by this time. What, however, it does contain in large quantities, is salts. Thus it is a dangerous article in summer diarrhœa, and must never be administered by itself, but in combination with farinacea, or raw albumen (which in this mixture, requires very little salt, if any).

Egg has been utilized as an admixture to milk, or as its substitute in a great many ways ; the yolk and the albumen have been so employed. The white of an egg, with a little salt, and six ounces of water, well beaten and shaken, is a good mixture, which, however, can take the place of infant food but temporarily, but is an invaluable make-shift in severe intestinal catarrh, or as a permanent nutriment in the same when added to other food.

Falkland, skimmed milk, and transformed it by means of pepsin, a process that does not recommend itself to general use by its circumstantiality; Roberts heated milk to nearly a boiling point, and treated with liquor pancreatis and bicarbonate of sodium; and the several methods of peptonizing milk are now generally understood and widely appreciated.

OTHER SUBSTITUTES.

Cow's milk, as a universal substitute for mother's milk, has lost credit with many through the differences in the article (which, however, as long as no adulterations are perpetrated, are less marked than those of the same secretion in the human being). If that were not so, how does it happen that all over the civilized world substitutes are sought for, offered, and purchased, though milk be as cheap and handy as anything else? Why is it that to avoid cow's milk, untold risks are run in procuring more expensive, more unknown, and more unreliable vegetable compositions, which seldom keep the promises loudly displayed on the labels? What are these promises? What does it mean when they claim to take the place of mother's milk?

Mother's milk is considered the best prepared article of food known. An ideal article of food must serve two purposes and consist of two classes of constituents. It must, in the infant, supply the growing tissue with material sufficient to take the place of that which is constantly wasted, and to allow a surplus for growth; and, secondly, supply fuel for the purpose of keeping up the production of an equable temperature, and the functions of the organs, mainly those of respiration. The first requirement is fullfilled by the proteinous, or albuminous substances, the other by carbo-hydrates. This statement would require modifications, if it were our object here to be absolutely correct, for the two classes will supply each other, act vicariously for each other, even change into each other, under certain circumstances, in the complex machinery of the human system. The albuminous tissue-builder in the milk is mainly caseine, sometimes, albumin; the second class is principally represented by fat and sugar. In vegetables the first class is represented in the gluten, the second mainly in the starch. If we add, that in milk, the ideal food, the proportion of the first class to the second is about 1 to 4, the vegetable substitutes are to be judged, in regard to their mere chemical composition, by the same formula.

But it is not the chemical formula alone which determines the rank of a substance as a nutriment. To the equivalents of the chemical formula of cow casein infant mathematics would not object, did not the infant stomach revolt against it. Thus it is not exclusively the chemistry of an article, but its digestibility, which comes in question. Now, not everything is equally digestible for everybody, sick or well, old or young, adult or child, infant or newly born; and particularly is this so with regard to starch, which forms such an overwhelming part in the composition of vegetables, and particularly of those which are mainly used for the purpose of the manufacture of infants' foods.

Starch is changed into sugar and rendered digestible by the secretion of two sets of glands, the salivary glands and the pancreas.

SALIVA.

The function of saliva is two-fold; first, to lubricate, and second, to transform starch into grape sugar. The latter change is also observed in plants. There is a large quantity of starch in the potato, with a very small proportion of the ferment that changes the starch slowly between spring and winter.

The ferment contained in saliva, which contains 99 per cent. water, acts in the same way.

The three pairs of glands which secrete it, begin to be developed in the second fœtal month, are quite noticeable in the third, and then develop rapidly. The parotids, for example, weigh two grammes (half a drachm) at the age of one month; that is, $\frac{1}{1500}$ part of the weight of the whole body, that is more in proportion than in the adult.

At the age of fifteen months they weigh five grammes, and eight grammes at the age of two years.

Since the time of Bidder and Schmidt, Ritter von Rittershain, Joerg, and myself in "Dentition and its Derangements," published in New York, in 1862, a great many experiments have been made with reference to the formation of sugar by the action of saliva.

Schiffer experimented upon babies at the age of two hours, sixteen days, and two months, and in every instance he found, as the result of the action of saliva on starch, sugar by Trommer's method.

Korowin made infusions of pancreas and of parotids, added starch, and the result was that the pancreatic infusion changed starch at a later period than did the infusion of the parotids. In his experiments the pancreas did not change starch in the first month, only slightly in the second, but noticeably in the third month.

The infusion of the parotids, however, was efficient in the first few days of life, particularly in infants of large size and well developed. The effect increased visibly towards the end of the first month, and the quantity of secretion increased to such an extent, that he could collect a cubic centimeter (fifteen minims) within five or seven minutes in the fourth month of life.

The saliva of seventeen babies, at the age of from one to ten days, exhibited the same diastatic power. A number of these babies remained under observation a long time, so that no mistake could be made. The number of his quantitative analyses amounted to one hundred and twenty. When he compared the diastatic effect of the saliva of a baby eleven months old with that produced by his own, he found that it was the same from the same quantity.

Since his first observations, Korowin has gone

over the same subject, and has given the results in the *Jahrbuch f. Kinderh.*, 1875, p. 381; they are as follows: It is possible to collect the secretion of the oral cavity in babies a few days old. Still there is some difficulty in gathering saliva before the age of six weeks. The quantity of this secretion increases towards the end of the second month, and this augments with every month. The secretion is almost always acid, unless the oral cavity is carefully cleansed. After it has been washed out, it is slightly acid, or slightly alkaline, or simply neutral.

From the very first month of life a distinct diastatic effect is produced by the oral secretion, and this increases with every month.

Infusions of the parotids, prepared at different times after death, produce the same effect. Infusions of the pancreas, within the first three weeks of life, have not produced any change; its diastatic effect begins with the fourth week, and remains feeble up to the end of the first year.

Zweifel has made a number of observations, and given the following conclusions. The infusion of the submaxillary glands of the infant does not transform starch into sugar even when it has been exposed to the influence of the infusion for one hour. The effect of the infusion of the parotid of a baby seven days old was distinct after four minutes exposure; that of the parotid of a baby that died of gastro-enteritis at the age of eighteen days, did not show itself until after the lapse of three-quarters of an hour.

There was no effect produced by a similar infusion made from the parotids of a child prematurely born, from one that died of diarrhœa and debility, from a fœtus in the third month, from a fœtus in the fourth month.

An infusion of the submaxillary glands of a fœtus in the ninth month of utero-gestation produced no effect upon starch. The parotids of the same fœtus produced a change after three-fourths of an hour.

It is a remarkable fact that different varieties of starch are not changed into grape-sugar in the same length of time. Solera found that the transformation of the starch of the potato was the most rapid; next was the starch of Indian corn, then that of wheat, and the change of the starch of rice was the slowest. The same results were obtained by Malay.

Raw starch changes slowly; boiled starch quickly. According to him, the starch of potato required from two to four hours; that of peas, from one and three-fourths to two hours; that of wheat, one-half to one hour; of barley, ten to fifteen minutes; of oats, five to seven minutes; of rye, three to six minutes; of potato paste, five minutes.

It is important to know that *the effect produced by the saliva persists in the stomach*, although this effect ceases within two hours.

It ceases altogether, and the starch will not be changed in the stomach as soon as the secretion of hydrochloric acid has begun in the digestive process.

This is a very important fact, because it shows that the *infant food, although it is not masticated and passes the mouth very rapidly, is still under the influence of the saliva in the stomach for some time.*

Hydrochloric acid is not secreted at once. The first acids in the stomach while digestion is going on, are of an organic nature, the lactic (and sometimes the butyric); thus it is that, when gastric juice is removed from the normal stomach, it contains organic acids only. So also, there is no free hydrochloric acid during digestion, for instance, in fever, a considerable amount of catarrh, or in dilatation of the stomach when the pylorus is constricted. In that condition amylacea are taken to advantage, principally because the diastatic effect of the saliva is not disturbed. In a gastrostomized boy, Uffelmann found that while there was no fever, there was lactic acid only in the stomach, and no hydrochloric acid during the first half-hour of digestion; afterwards, hydrochloric acid was found.

Some starch is digested at the very earliest age. If there be a moderate surplus, it is expelled like the surplus fat in normal woman's milk, without annoyance or injury. Besides being nutritious, to a certain degree and in its peculiar way it serves to dilute cow's milk, to reduce its percentage in casein, to prevent the latter from coagulating in large masses, and thus to render it digestible. To accomplish all this, no large quantity is required. Thus those cereals and farinacea are to be preferred which contain a small proportion of starch and a large one of protein, or those substances (gum-arabic, gelatin) which, while serving the above indications, are also nutritious. Of cereals belonging to the former class, I prefer barley and oatmeal. Thus the number of available articles is by no means small; and they all come up to the requirements we look for in such substances, viz.: *They are accessible, and for sale everywhere; the mode of their preparation is perfectly simple and easy; and they are cheap.*

These requirements are not always fulfilled by the artificial foods offered for sale; I cannot help referring to them again, though my so doing before has not increased the number of my friends amongst the advertising manufacturers.

The community insists, with the utmost pertinacity, upon giving their babies, as soon as weaning-time arrives, or before, articles of food such as they know nothing about. When an adult sits down to a meal and finds placed before him articles of food with which he is not familiar, he makes inquiries regarding them before eating. The baby, however, is credulously fed upon things with which the child, father, mother, or doctor, has not the least familiarity—foods which are sold in large quantities, but have a composition unknown to the public. When some manufacturers deign to say anything about their merchandise, it is to the effect that the food offered is the best in the market; that it is the proper thing and the only thing for children and invalids of all ages; that the relation of albuminous substances to carbohydrates is exactly correct; and that a package costs a certain amount of money. In regard to this subject the public appear to be smitten with absolute blindness. They insist upon forgetting that the man who offers for sale, and advertises at a very heavy expense, does so, as society is constituted, for his own pecuniary advantage solely. To say that when the article offered is not good it will find no market, is deceiving ourselves, experimenting on our babies, relying on the character of a single man or corporation, on the honesty or intelligence of the manufacturer's chemist, superintendent—on the nature and condition of the elements used in the composition of the article—and on ever so many other influences, which can work before the manufactured article gets into the hands of the consumer. Why the sellers and advertisers of unknown compounds should be more trusted than those who sell a simple article of food, such as milk, which is constantly adulterated, can hardly be perceived. Is it necessary to say that the factory furnace is always lighted more in the interest of the proprietor than for the benefit of the public!

Still, in regard to the growing evil, which has assumed vast proportions, the profession is to a certain extent at fault. There are few but are aware of the inexpediency (and sometimes danger) attending the exclusive feeding of cow's milk, and consequently they look for substitutes. Examples of infants thriving on almost any food are numerous; the public taste runs in the direction of the unknown; thus the responsiblility of advice or assent is but a slight one; many of the foods in the market come in a pleasant form and convenient for use; thus the food business firm thrives. Professional men have come to look upon the use of patent foods as something quite unobjectionable. Those imbued with the strictest sense of ethics, who would not patent an invention, nor tolerate the fellowship of a professional man who does so, who frown upon patented medicines because they are unknown and unknowable compounds, even though their components be printed on the labels, these very men forget their habits and principles when the question of patent-right and secrecy comes up in regard to foods. If I add, that many of the scientific journals of Europe, particularly those of Germany, dedicated to the study of children's diseases, are frequently used for the purpose of discussing the merits and effects of some new infants' foods, it is only to show to what extent the evil has grown.

No profound thinking is required to appreciate the fact that of a great many of the articles offered for sale, a few only are available compositions. But the very fact that they are compositions, that everything organic may spoil, that every compound depends on too many circnmstances which are apt to interfere with its uniform condition, and that when we rely on a compound, we rely at the same time on a proprietor, his foreman, his workman, his chemist, and the wholesale or retail dealer, we feel that we are easily deceived or disappointed. Besides, for an article the constituents of which we can purchase at a low price, we are taxed to an inordinate extent. I repeat what I often said before: Artificial foods must be simple, recognizable, accessible, cheap, and easy to prepare. Thus only will they become universally useful to the rich and poor, to city, country, prairie and backwoods.

BARLEY AND OATMEAL.

Oat-meal has been a preferred article, and has long enjoyed the recommendation of authors as a proper substance for children's food. Van Swieten praises it in exceptionally high terms, and T. Herbert Barker placed it at the head of articles of diet twenty years ago. In placing oat meal gruel at the head of the list of farinaceous foods, I am guided by my own observation of its utility. Such, indeed, is my confidence in its value, that if I were restricted to the use of any one article in addition to milk for bringing up a child, it should be this and barley-meal.

I have always preferred barley-meal, when an article for steady diet was to be recommended, for the reason that oat-meal, on account of its containing fat and mucin, tends more to relax the intestines;

otherwise, the chemical composition of both is so nearly alike, that it would make but little difference upon which the choice should fall. Meanwhile, there is no danger so common for little children as that which occurs on account of their tendency to diarrhœa. My advice in this matter, once for all, is to give barley-meal to children who have a tendency to diarrhœa, and oat-meal to those who have a tendency to constipation, and to substitute occasionally the one for the other as changes occur in the performance of the functions of the intestinal tract. One more observation may be made at this juncture on account of its practical importance, namely, that diarrhœa and a milk diet are not compatible, therefore, when barley meal must be given on account of diarrhœa, it is advisable to lessen the quantity of milk at once, or suspend it entirely for a time. In the latter case the white of egg, either with or without brandy, must take the place of milk. This plan has led me safely over many a danger in the past thirty years, and has saved the lives of many children. It has given me great pleasure to see the success which K. Demme (Dreizehnter Jahresbericht, 1876), Hennig, and others since, have had by following a similar plan. In my book upon " Infant Diet " I made the assertion that it is a matter of indifference whether you use for children the barley corns with the husk retained, or the smaller pearl barley with the husk removed. That statement is erroneous. I had been misled by the supposition that protein bodies and starch are uniformly distributed in barley; in point of fact, however, the relation is such as is found in the other grains, that is to say, by far the greater part of the gluten is collected in and under the superficial coverings. According to Enzinger's publications, the meal granule of the barley-corn, which lies next to the layer covering the seed, is composed of large irregular cells which are filled with albuminoid corpuscles, and yield no starch.

Further inward are found large irregular quadrangular cells, which yield albumen and an abundance of starch. Still further in the interior are yet larger cells which are filled almost exclusively with starch. The consequence of all this is that the entire kernel, and not the interior alone, is to be used for food for children. That which is found in the shops as "prepared barley-flour" is fine and white,—two suspicious qualities. The less the meal possesses of the yellow outer layer, which contains gluten, the less reliable is it; likewise, the price of this refined article is fixed so high that the pretense of purification is paid for at a good round sum. I I advise, therefore, in all cases, that people grind the barley to be used for children's food as finely as possible in a coffee-mill reserved for that purpose, in order to shorten the time necessary for boiling, and thus save the gluten. In the case of very small children it will pay well to allow the barley to boil for hours, in order to let the outer layers burst and void their contents, as in this way the largest part of the starchy matters in the center will be got rid of by straining the decoction. It is always a good plan to grind the kernels as already recommended, and not to make use of the pearl barley, which is the inner kernel freed from the husk.

For children who have passed the first few months of life, the pearl barley answers fairly well, with its large quantity of starch. It remains to be remarked that in large cities the simplest crude materials are only to be had by special inquiry for them. Now, by the addition of barley or oat meal to the milk which has been previously prepared in a proper manner, I expect more than a mechanical dilution, because when Moleschott declares that thirty-six ounces of barley meal are sufficient for the daily fare of a full-grown laborer, the addition of from ten to twenty-five grammes (two to six drachms) of the same material means no insignificant increase in the diet of a child. C. Voit has recently made the needs of children in the way of nourishment at particular periods of life, the subject of a searching investigation (üb. d. Kost in öffentl. Anstalten, Z. f. Biol., xii, 1, 1876). More recently Sèixler (Ernährungsbilanz der Schweiz, p. 6) has calculated the relation of the ingredients of food proper for children up to the age of fifteen years, with the result that the following allowance should be made: Albumin, 78 gm.; fat, 20 gm.; carbohydrates, 250 gm.;—nitrogenous substances: carbo-hydrates, 1:3.8.

Hildesheim calculated for children of from six to ten years: Albumin, 69 gm.; Fat, 21 gm.; Carbo-hydrates, 210 gm. (1:3.6).

In the orphan asylum at Munich, where the children present a good appearance upon a mixed intellectual and physical supply, Voit calculated for their diet: Albumin, 79 gm.; Fat, 35 gm.; Carbo-hydrates, 251 gm. (1:3.9). Finally, he made a careful calculation in which the requirements of a child are compared with those of a grown person, at rest or at work. A child of 10 to 11 years, weighing 23 kilos (46 pounds): Albumin, 79 gm.; Fat, 35 gm.; Carbo-hydrates, 251 gm. (1:3.9).

The laborer of 60 kilos (125 pounds) averages: Albumin, 118 gm.; Fat, 56 gm.; Carbo-hydrates, 500 gm. (1:5).

The same laborer, while at work: Albumin, 173 gm.; Fat, 173 gm.; Carbo-hydrates, 352 gm. (1:4.7).

During rest:

	Alb.	Fat.	Carb-hyd's.
1000 kilos in a child.....	343	152	1091
1000 kilos in an adult...	228	120	586

Thus it follows that a child of a given weight, who, at the same time must keep up his tissue changes, and must supply albumin, fat, and ash residues, requires absolutely more of the important elements than the adult in the same condition of comparative rest. Albumin to the extent of 50 per cent., fat 25 per cent., carbo-hydrates about 100 per cent., are demanded by the child, under the above circumstances, in excess of that required by an adult. The substances containing these elements must not only be received, but they must be taken in digestible and assimilable form, and in the same or nearly the same proportion which they hold in the mother's milk, in which the relation of the nitrogenous to the non-nitrogenous elements is as 1:2.9. Where it is possible this relation must be sustained.

J. Foster has given the following table of relative requisites:

Age.	Food.	Alb.	Fat.	Carb-h.	Proportion.
7 weeks.	Gruel	29	19	120	1:5.3
4-5 months.	Milk	21	18	98	1:6.1
1½ years.	Mixed	36	27	150	1:5.4

Without doubt the above proportion of 1:6.1 is to be attributed to the considerable quantity of sugar in the milk. To what point this element is useful, indifferent, or harmful, has already been alluded to, in part, but must be still further elucidated. It will be shown that under certain circumstances it may be useful, is occasionally harmful, and seldom quite indifferent. It is too soluble, too readily absorbed, too easily changed to be quite indifferent. In this respect it is quite different from a small quantity of starch, which may pass through the intestines without being changed or working any appreciable changes. Indeed the intestines of children have to pass every day some of the natural food, that is, mother's milk, or some parts of it (principally fat) without any change.

Still I emphasize the fact alluded to in my explanation of the functions of the salivary glands, that additions to milk diet during the first few months of life must not contain too large a percentage of starch, while at a later period it is not necessary to be over-particular in regard to the same. I have therefore stated that the barley-gruel which is given to very young children, should be prepared from the crude barley-corns, because of the greater quantity of protein material that is contained in their outer layer.

E. Wolff has called attention to an important fact in reference to the nourishment of animals used for labor upon the farms, viz., that the feeding of an easily digestible protein results in no real change in the digestible properties of the rest of the food, but that a considerable addition of carbo-hydrates accomplishes a greater or less depressing effect in the digestion of the other food.

Thus, if we are justified in using the results of these experiments upon animals, the quantity of starchy material which is to be mixed with the food must not be too great, nor is it proper to select farinacea at random; in this respect many mistakes have been made even by eminent writers. For example, while I can only approve of Soltmann's recommendation to use gum arabic for the purpose I have mentioned, I am compelled to doubt Pepper's recommendation of gelatin, and arrow-root, in order to facilitate the separation of the coagulated casein; they are by no means equivalent. Just as little can I accord with the recommendations of Eustace Smith who employs for the same purpose indifferently, arrow-root, sturgeon's bladder, or gelatin substances which are quite unlike each other; the bad results which he must have seen from this lack of discrimination, form very good reasons why he should recommend the frequent use of carminatives. In "The Sanitary Care of Children and Their Diseases" (a series of five essays by Drs. Eliz. Garret-Anderson, S. C. Busey, A. Jacobi, J. Forsyth Meigs, and J. Lewis Smith, prepared by request of the trustees of the Thomas Wilson Sanitarium of Baltimore, Md.," Boston, 1881), Doctor Busey says "Notwithstanding the protest of Jacobi against the use of rice-water, the experience of the writer coincides with that of Trousseau in regard to its value; we have too often witnessed its beneficial effects in young infants, in cases of diarrhœa with uncontrollable vomiting, to abandon its employment on mere theoretical grounds." This objection is not to the point, for I am not now, nor was I then, dealing with pathological conditions and therapeutical indications, but only with the consideration of farinaceous foods under normal circumstances, and for the purpose of a regular diet.

GUM ARABIC AND GELATIN AS ADDITIONS TO MILK.

Until within a short time gum arabic was not considered nutritious. Frerichs, Lehmann, Husemann, will not admit that any change takes place in it; Gorup-Besaney says that it is dissolved but not digested; mixed with milk it is presumed to act only in a mechanical manner. Some years ago, however, Uffelmann experimented upon a boy with a gastric fistula, with a solution containing eighteen parts of gum in two hundred parts of water, and made direct observations. This solution, introduced directly into the stomach, yielded after a while, without the presence of saliva, a trace of grape sugar; fifteen grammes of the gum yielded in forty-five minutes a twentieth part of a gramme of grape sugar; thirty grammes in sixteen minutes yielded twenty-eight one-hundredths of a gramme (0.28) of grape sugar, the latter with quite an acid reaction, though it was not decided whether the acid had already existed for some time in the stomach, or was only developed with the introduction of the gum. In both cases the presence of hydrochloric acid and conversion into grape sugar seem to go hand in hand. Probably the addition of some hydrochloric acid will recommend itself in practice, if the object is not only to obtain the mechanical effect of the gum, but also its final changes. Milk, also, when it is to be administered with the gum arabic, bears the admixture of muriatic acid quite well, and it will be remembered what has already been said of J.

Rudisch's plan to render milk more digestible by mixing with water and this acid.

In reference to gelatin there are two indications. Some lay most stress on its mechanical action in distributing the elements of the milk in the same way that this is brought about by additions of gum and farinacea; others insist upon probable usefulness in effecting tissue changes.

Guérard compared a large number of reports made for the French Academy, and others, and came to the conclusion that gelatin is very nourishing, in the first place; secondly, that on account of its probable importance for the cellular tissue it is absolutely necessary for the maintenance of life. He quotes Jean de Sery as follows: "Having tried what skins and parchment are fit for in case of necessity, if I had buff jerkins, clothes of chamois leather, and such other things as contain sugar and moisture, and were confined in a fortress for a good cause, I should not desire to escape from fear of hunger." In the same way Denis Papin is said to have made an offer to Charles II (La manière d'amollir les os. Paris, 1682) to prepare for him, out of bones, for use in poorhouses and hospitals, "a quintal and a half of gelatin, by means of eleven pounds of charcoal." In a similar manner D'Arcet promised to make "five oxen out of four", by utilizing the gelatin.

Voit experimented for nine days on a dog which was fed upon gelatin. He found conclusively that gelatin does not build up the body, and is not, as Guérard supposes, deposited as a gelatin-yielding tissue, but is readily decomposed and broken up in the place of the circulating albumen, which is thereby protected and saved. In this way it saves the circulating as well as the organized albumin, and accomplishes the same results as the carbo-hydrates and fats, only to a more intense degree. In opposition to these assertions of Voit, Tatarinoff concludes that it is never possible to prevent a diminution of the weight of the body when an exclusive diet of gelatin is administered. The quantity of nitrogen which was discharged in the form of urea, in the experiments made by him, was always greater than that which existed in the gelatin consumed; the loss of weight during a gelatin diet was materially less, however, than in complete starvation, when diarrhœa and bloody urination could be avoided. When starch, fat, extract of meat, and water, were combined with meat or gelatin, it could be shown that the weight of the body increased with the former and diminished with the latter; furthermore, the quantity of urea which was discharged was much less with the addition of meat than with the addition of gelatin; the nitrogen in the urea, after the addition of gelatin, was absolutely more in volume than that contained in the gelatin itself while the meat contained a greater volume of nitrogen than was discharged. Hence it appears that the nutritious value of gelatin is much inferior to that of meat, but the addition of gelatin results in smaller loss to the body in a mixed diet of starch, fat, meat-extract and water, than occurrs without such addition. Likewise it was evident that the other ingesta were more completely digested, and that a smaller quantity of excrementious matter was discharged when gelatin was added than when omitted.

Gastric juice (but not the acid portion alone) so alters gelatin that it no longer gelatinizes. The whole question, however, appears to have been finally answered by Uffelmann. Upon his patient with gastric fistula, first when in a condition of fever, then without fever, he found by direct observation that gelatin is actually dissolved; within an hour it is so modified that it no longer holds together, and is easily diffused. Artificial gastric juice, however, required from eighteen to twenty hours to accomplish the same end. No bad odor was developed any more than in ordinary stomach digestion. However, grape sugar, which is occasionally found in ordinary gastric digestion of gelatin, was not observed. Thus, gelatin it easily digested, but readily decomposed, and requires the addition of hydrochloric acid. This latter observation is of great practical importance, for in acute and debilitating diseases the secretion of pepsin and of hydrochloric acid is diminished. There is no doubt, therefore, as to the utility of gum-arabic and gelatin as an addition to cow's milk and to children's diet. Not only do they fulfill the indication of dimin-

* 6

ishing and distributing the particles in cow's milk, but they also officiate as a means of direct nourishment by preventing waste. Furthermore, they are satisfactory articles, inasmuch as they are simple materials which can be obtained everywhere, are out of reach of the patent medicine line, and are cheap, and easily handled — simple boiling suffices to produce a solution.

HOW TO FEED.

Should little children be fed from a spoon, a cup, or nursing bottle?

Most certainly from the latter. It alone gives certainty that the food has suitable consistency and contains no lumpy ingredients. The accurate removal of lumps, and a uniform consistency of the food, in the child is analagous to mastication in the adult—at least, this is approximately true: the prejudice which is prevalent with mothers and many nurses, that thick nourishment is necessarily nutritious, must be met and opposed energetically. Proper digestion demands, above all, a gradual introduction of food into the stomach to which the gradual secretion of the gastric juice must correspond. The use of the bottle is so much more indicated, as well as desirable, in that, when a slight degree of weariness comes on after its use, the infant is naturally obliged to cease nursing; especially is this an important point when cow's milk and other thickened nourishment is used. A common clinical experience, even with grown people, is that they will not endure milk so long as they drink it rapidly, but that they have no trouble when they can take it with a spoon, slowly. Besides, the act of sucking, in itself, excites peristaltic action, and secretion of the digestive fluids (Spallanzani, Brown-Séquard). The digestive tract is a continuous canal, the sucking action excites the function of the salivary glands, and arouses that of other glands. When Th. Ballard wrote his book on the diseases of women and children, thirty years ago, in which he proposed to show that almost all the diseases of children, and a good part of those of women, were the result of fruitless sucking at empty breasts or empty bottles, he was naturally laughed at for his over-statements, whereas he was influenced by physiological hypothesis and clinical experience.

In reference to the handling and cleansing of nursing bottles, the greatest care must be observed. Even before the bottle is filled, the nourishment, especially if it be milk, is exposed to decomposing changes. The residue which remains in the bottle and upon the mouth-piece, especially if composed of rubber, ferments very quickly, and for that reason is apt to be dangerous. There is but little difference as to the kind of nursing-bottle used; if people are disposed to be painstaking and cleanly, even complicated nursing-bottles will be properly cared for, but those who are careless will evince their disposition with even the simplest form of apparatus. The kind recommended has a rubber tube from 16 to 20 centimeters long between the bottle and the mouth-piece, and also a glass tube united to the rubber and extending nearly to the bottom of the bottle.

There are also " biberons pompes." In these a hollow rubber ball is introduced low down in the glass tube whose end projecting out of the tube is converted into a valve by an oblique slit which is made through half of the ball. By simple pressure upon the mouthpiece with the finger, the current appears in jets. Since I gave the first description of the apparatus, Soltmann has made improvements in it, and I have made it accessible to American readers through an article in Buck's "Hygiene" (New York, 1879). In many cases it has done good service. As a rule no difficulty in sucking can come to a healthy and well developed child, yet there are causes of many kinds which may render it troublesome or impossible. Immature and sickly babies, or those with cleft palate do quite well with this self-acting nursing bottle, from which the contents are sent into the mouth by the very slightest pressure on the mouth piece.

How often ought a child to get nourishment? The proper number of times for nursing is very differently estimated. Natalis Guillot, who introduced the method of systematical *weighing* for the purpose of determining the influence of nourishment upon weight, advised 20 to 25 nursings daily of perhaps 25 grammes breast milk each time. Bouchaud reduced this estimate to 8 or 10—On the first day of life each nursing was to consist of three grammes of milk, on the second of 15, on the third of 40, on the fourth of 55; according to his opinion, the first day called for 30 grammes, the second for 180, the third for 450, the fourth for 550, every day of the second month for 650, of the fourth for 750, of the fifth for 850, of the sixth until the ninth for 950 grammes. Ahlfeld allows five nursings daily to a child from four to eight weeks of age, afterward five or four. Fleischmann approves of ten or eleven. According to my experience the newborn infant should be fed from eight to ten times daily; after it is four months old, five daily nursings will suffice.

Practically it is not very difficult to regulate the quantity of nourishment taken. Healthy children will fix the proper limits themselves. The same rule will apply, with slight differences, in an equally good condition of health, for nourishment artificially prepared. But it is impossible to lay down as a uniform rule, that some particular quantity must be considered as normal for all children. Ahlfeld calculated 104 grammes at a nursing for the fourth week (the minimum being 50, the maximum 140), in the tenth week 164 (min. 110, max. 225), in the twentieth 212 (min. 100, max. 325), in the thirtieth 263 (min. 85, max. 350).

All such calculations count merely as experiences, not as infallible rules. The child ought to drink from the breast or the bottle until it has had enough. It takes 20 to 25 minutes to empty one or both breasts, and having nursed, the infant should be quiet, should play with its arms, should breathe somewhat more regularly than usual, or go to sleep. If it be not allowed to have absolute rest, if it be rocked, allowed to lie on its belly, or be carried around lying face downward upon the hand, vomiting will be provoked. The facts which have been introduced show that Nature requires *play-room*, and that even the most careful measurements may be incorrect. For instance, Fleischmann found on carefully inflating the stomach of a child four weeks old under a pressure of fourteen centimetres (having dried and varnished it), that its capacity was eighty cubic centimetres. He also treated the stomach of a child two months old in the same manner, and found its capacity to be one hundred and forty cubic centimetres. "Accordingly the quantity of milk to be taken at each nursing must amount to from eighty to one hundred and forty grammes." In the same way he found the capacity at the seventh and ninth months to be one hundred and sixty and one hundred and eighty grammes. Is there anything which can be more "accurate and correct?" On the other hand, Ahlfeld found by weighing, that the proper quantities were 200 to 210 grammes, instead of 140, and from 200 to 300, (even 350 to 400) instead of 160 to 180 grammes, and concluded that Fleischmann had either made a mistake or had treated the subject to no purpose. Can anything be more "accurate" than weighing or more absolute than the result so obtained? Both investigators were right in their direct results, and wrong in their conclusions. Both forgot that the nursing, which occupies a child 15 or 30 minutes, represents no invariable quantity, and that the whole quantity cannot be found, at any time, entirely in the stomach. Apart from the elasticity of the living organ in comparison with the dried and prepared one, it must be borne in mind that rapid absorption takes place in the former from the moment of the reception of food; hence, an infant with an apparent stomach-capacity of 80 to 140 cubic centimetres can just as readily take up 200 grammes of fluid food, as an adult can dispose of 1,000 grammes within half an hour when he so desires.

NOURISHMENT SUCCEEDING THE PERIOD OF INFANCY.

In the course of the second half year some changes may be made in infants' diet. In the relation of barley preparation to milk, the latter may exceed its former quantity, and in the same proportion in which the children are permitted and accustomed to drink pure water, the food may become more condensed. Towards the end of the first year, the quantity of barley or oat-meal to be used in the decoction may be increased. It is soon enough to begin the use of pure milk in the third half year, if at all. About the eighth or the tenth month the chewing of a crust of bread, or of a piece of " zwieback " may be allowed; about this time, too, the daily allowance of meat soup may be increased to 250 grammes, and in addition one or two teaspoonfuls of broiled beef may be given. These articles distributed through four, or perhaps five meals, will be sufficient for the greater part of the second year. The quantity may be gradually increased, but more radical change is useless. If a child which is healthy and not spoiled awakens at night, it needs and desires nothing but a drink of water, or of thin barley water without milk.

About the middle of the second year when the child begins to use a spoon, the breakfast may be made up of more solid elements than heretofore—barley broth or oatmeal mush thoroughly cooked, an egg, a glass of milk, a piece of stale bread with or without butter. The child must be taught never to drink milk in haste; it will be digested better when time is taken. The daily quantity of meat, preferably beef, to which gradually may be added lamb or chicken, may now be increased to 100 grammes, and this is to be at two or three meals. The evening meal must be similar to that of the morning, and lighter than the mid-day meal. Neither at this age, nor later, should nervines, stimulants, condiments, coarse vegetables or salads, coffee, tea, wine, beer, play any part in children's diet. A piece of sugar after a meal which is frugal but rich in albuminoids, will prove an agreeable and useful addition. Children from two to three years of age, will get along well on four meals daily. Those who are a little older, may do with three, provided they get once a day between meals a piece of bread, and a drink of milk made agreeable and more digestible by the addition of a little salt. Before children are two years of age no vegetables *in any quantity* should be given them. Small quantities may be given later on, when they will be acceptable and readily digested. As age advances, the diet should approximate, more and more, that of grown people. Altogether there is no easier, and no more grateful task than that which consists in accustoming children to a simple diet, and to shapeing their habits and their demands in harmony with those of Nature, from the first year of life. Thus only a prosperous development, both moral and physical, can be expected. C. v Voit (Unters. d. Kost. in einigen öffentl, Anstalten in München, 1877) estimates the daily quantity of food required, as: Albumin, 79 grammes; fats, 37 grammes; carbo. h. 247=nitrogenous: non-nitrogenous=1:4.

J. Foster estimated for a well nourished child of one and a half years: Albumin, 36 grammes; fats, 37 grammes; carbo. h. 151=nitrogenous: non-nitrogenous=1:5.4.

Th. Riedel estimated for a public institution at Berlin: Albumin, 74; fats, 18; carbo-h. 433=nitrogenous: non-nitrogenous=1:6.3. This resulted in a weakness of muscular structure throughout.

Koenig proposes for children of 6-7 years the following:

	GRAMMES.	ALBUMIN.	FATS.	CARBO-H.
Raw meat	170	30	1.2	0.
Bread	300	19.5	1.	150.
Potatoes	180	3.	0.4	36.
Fat	25	....	25.	....
Milk	250	8.5	9.	12.
Flour	100	10.	1.	74.
Vegetables	180	7.	1.	9.
		78.	38.5	281.

The foregoing corresponds also with the following bill of fare: Raw meat, 100 grammes; cheese, 2.5; bread, 300; potatoes, 180; fat, 20; milk, 250; flour, 100; vegetables, 180. Or the following:

Eggs, $2 = 100$ grammes; peas or beans, 100; bread, 250; potatoes, 180; fat, 25; flour, 100; vegetables, 180; milk, 150; beer, 150; and in addition: coffee, tea, and condiments which contain but a minimum of nutritious substances.

THE MOUTH.

The general nutrition of infants is affected considerably by malformations of the mouth. Most of them are very liable to render nursing difficult, and may prevent it altogether. Sometimes babes are so feeble that it is impossible for them to suck their own mother, especially when she is a primipara, though they may be able to take the breast of a wet-nurse who is a multipara.

Small hare-lip, when uncomplicated with fissure of the palate or maxillary process, prevents sucking only when the babe is very feeble. Hare-lips which are complicated with fissure of the palate always prevent sucking. An undue length of the soft palate does not prevent sucking to such an extent as does shortness; when it is too short a vacuum cannot be formed.—Once I saw a soft palate in an idiot, which was immovable and transparent; it lacked its muscular layer entirely, and deglutition and articulation were very deficient. For very many months efforts were made to teach this boy to articulate, and this without once examining the mouth.

Small defects in the hard palate and large defects in the soft palate are impediments to sucking.

Along the median raphe of the palate there are in almost every new-born child, small dots, white, elevated, and found in clusters. They have been described by Bohn as sebaceous follicles, but found by Epstein to consist of nothing else but accumulations of epithelium. I have not seen them ulcerated at birth, but after they had lasted for some time and had been either neglected or maltreated, ulcerations sometimes formed, which perhaps reached down to and into the bone; and I have seen these small (originally physiological) deposits give rise to difficulties in regard to sucking.

Cohesion of the lips in the median line, as are sometimes described, I have not seen; but fissure of the cheeks frequently occur and generally impeded sucking.

The tongue, however, is at some time or other the cause of the inability to suck, no matter whether there are fissures in it or whether there is a fibrous or lymphangiectatic hypertrophy. This so-called makroglossia is the more serious, the more liable it is to grow, and as a rule it is connected with other deformities, and in many cases with idiocy. Congenital sarcoma of the tongue I have described in the *American Journal of Obstetrics*, Aug., 1889. Vascular tumors are not infrequent.

The frenulum of the tongue has been accused of

exerting a very great influence on sucking. It has been operated upon with either knife or scissors, or has been torn with the finger nail, and often without cause. When there is the slightest mobility of the tongue forward and backwards, there is certainly no impediment to sucking, because it is always possible for the baby to produce a vacuum in its mouth. Now and then difficulty of articulation will be found in later life, when it is time enough to operate because of that difficulty; otherwise I have observed no ill effects from shortening or *elongation of the frænulum*. Still there are a few cases in which by elongation of the frœnulum, the tongue becomes so moveable as to enable the baby to double it up, thereby producing difficulty in respiration; there are even a few cases in which what has been called "*swallowing the tongue*" has taken place, when the tongue was doubled backwards, sucked into the larynx and gave rise to immediate suffocation. Such cases are reported by Petit, Levret, Hennig, and others.

A serious impediment to sucking, and therefore to general nutrition, is *thrush*, which as a rule, occurs a few days or weeks after birth, though it may occur later; it can be serious enough to weaken even a healthy child, and thus become dangerous, and even fatal, although thrush of the œsophagus and stomach are very rare indeed.

THE TONGUE.

The tongue of the newly-born is generally rather whitish, and remains so a number of weeks. It is very liable to participate in the diseases of the mouth, and its epithelium is easily changed and thrown off. Hence it is that it reddens on the slightest irritation, and gives rise to more or less superficial, longitudinal or transverse, mostly quite innocent fissures. This is the case, however, more particularly in atrophic children, or those who have suffered from gastric catarrh for some time.

There are a number of changes belonging to the tongue in the infectious diseases. That of scarlatina and measles is well known, and a similar condition is now and then found without apparent cause. What has been called psoriasis of the tongue is, in fact, no disease, but merely an anomaly of the epithelium which, easily thrown off, shows itself in white islands, circles, and semi-circles. This can remain without producing the slightest influence for many months.

The tongue also participates in the general condition of the mouth in thrush and all other forms of stomatitis, *diphtheria*, and the inflammation secondary trauma and combustion (mostly by hot fluids.) Superficial changes are also seen in diseases of the stomach and intestines; as a rule, however, the fur is very light, remains whitish in color, and is seldom very thick. Thus it follows that the tongue is very much more liable to be the seat of local affections, and that it participates in the affections of the oral cavity in general, but is not so liable to serve as an indicator in gastric and intestinal diseases as in the adult.

The frequency of "muguet," or thrush of the vagina in pregnant women, renders the early appearance of the same affection in the mouth of the babe a natural occurence; besides, the *oidium albicans*, characteristic of thrush, is identical with the *oidium lactis* which accompanies acid fermentation of milk. These two forms of thrush cannot be distinctly diagnosticated from each other.

To prevent, or remove it, lotions of cold water, or an alkaline solution, after every meal, after every vomiting, sometimes every hour, or even oftener, will suffice. It must not be forgotten to wash the nipple also with an alkaline solution after each nursing, for drops of milk remaining upon it give rise to fermentation and local irritation through deposits of bacteria and vibriones in the fissures below the surface, from which the mouth of the child may again be infected.

Neither should the floor of the mouth be left unexamined, for *ranula* is not infrequent, or similar cystic tumors, sometimes in the median line, sometimes laterally. No matter on what cause they depend, they are very apt, by pushing the tongue upwards, to render it immovable, and sucking impossible.

DENTITION.

The formation of the teeth begins in the first third of embryonic life. According to Goodsir, narrow grooves are formed in the sixth week of utero-gestation between what is afterwards to be the lips and the rudimentary maxillary processes, at a time when the former are hardly visible. The first change consists in the formation of wart-like excresences upon the bases of the grooves, the latter, as it were, forming receptacles for these excrescences. This is the first indication of the dental sac with a dental papilla in its cavity. In this cavity the dental substance is gradually deposited.

This is the way in which the dental sacs of the twenty milk teeth are formed. They undergo ossification in the fifth month of utero-gestation. Behind them are the sacs for the permanent teeth, but whether or not there is a communication between those of the former and the latter, is not yet known. After they have been separated from each other, however, there is still some connection between them and the "gubernaculum dentis." The separation is complete when the fœtus is finally born. About this time the margin of the dental cavity is cartilaginous, and the root of the tooth begins to grow, and by its formation and growth the tooth itself is pressed forward. During this process the cartilage of the wall and the gums is made to disappear. The lateral wall of the dental sac becomes the periosteum of the dental root; sometimes the cartilage disappears before the tooth has reached it, in which case the tooth can be felt before it can be seen. The two lower incisors will appear, as a rule, between the seventh and ninth month, when there is an intermission of from three to nine weeks, the upper incisors appearing between the eighth and tenth month, with an intermission following of from six to twelve weeks. Six more teeth make their appearance between the twelfth and fifteenth months; that is, two upper molars, two lateral lower incisors, and two lower molars. This growth is followed by an intermission, of from three to six weeks or more, when four bicuspids protrude betwen the eighteenth and twenty-fourth months, and the four second molars between the twentieth and thirtieth months.

The second dentition begins with the protrusion of the third molars, and this takes place in the fifth or sixth year. About that time the arteries of the temporary dentition are obliterated, the nerves disappear, the alveoli become large, and the teeth fall out without any caries taking place. At that time the temporary canine lies in front of the external incisors and the first molar. Thus it is that very often in later life the teeth have an oblique position. The wall between the alveoli of the temporary and the permanent teeth becomes slowly absorbed and the milk teeth fall out painlessly, unless the roots of the teeth have not been absorbed in the order of their first appearance.

In the twelfth year there protrude four more

molars; between the sixteenth and the twenty-fourth four more, the "wisdom teeth," the crowns of which ossify as late as the tenth year of life.

There may be great anomalies with regard to the appearance of the teeth. Now and then they have been found at birth, and then generally the incisors. Some of them hang loose in the gums; some, however, are solidly imbedded in the gums. Such occurrences are rare, hence in some parts of Germany and Switzerland a child born with teeth was formerly regarded as a witch. According to the missionary Endemann, some Asiatic nations throw a baby with congenital teeth or other malformation into boiling water.

About some there is a tendency to development of pseudoplasms. *Maxillary cysts* are mostly congenital, and are either follicular (that is, the result of dilated dental sacs), or periosteal, originating chiefly in the periosteum of the teeth and not of the maxilla. These cysts may contain bones and teeth. Latterly they have been explained by proliferation of embryonal cells, or have been regarded as duplicatures of the external embryonic layer.

Other congenital malformations are *cystomata, myxomata, sarcomata, fibromata*, which originate during the embryonic growth of the pulp of the teeth.

Aberrations from the normal time of the appearance of the teeth as given above are not rare. Sometimes they will come too soon; sometimes very late, for instance in rhachitis; at the same time the fontanels will close later than the normal period of fifteen months, the development of the bones of extremities is also delayed, and the lower jaw is small. Thus it is that, after a while, when the permanent teeth are expected, they crowd each other and become irregular. Rhachitis not infrequently develops during fœtal life, and then, sometimes, several dental sacs are merged into each other, and instead of two teeth we have only one; this is a frequent occurrence with regard to the lower incisors, and corresponds with the insufficient development of the lower maxilla in rhachitis. Teeth will also appear at a later period than normal when the children suffer from chronic disorders, such as anæmia, slow convalescence, etc.

The protrusion of teeth may be premature. When this occurs with syphilis or rhachitis, as a 'rule there will be a long interval after the first have appeared before those of the next growth announce themselves, say from four to six months. Generally, however, premature appearance of the teeth is connected with premature ossification of the bony system in general, and of the fontanels and sutures of the cranium in particular. When this is the case, the upper incisors usually appear first, undoubtedly in connection with the fact of premature ossification of the upper part of the cranium. This is a serious occurrence. When premature ossification is congenital, it makes parturition difficult and renders the child idiotic or epileptic; and it will have the same influence when it occurs at the age of three or four months; it will also exert a moderate influence of the same kind when it occurs from the eighth to the tenth month. At all events, it is impossible for the brain to develop favorably when its bony capsule does not permit of proper expansion.

It is a peculiar fact that even savage nations have made observations which show their fear of such an occurrence: the Makalaka in South Africa are in the habit of observing whether or not the upper teeth come first; in Bohemia it is a popular belief that the child whose upper incisors come first will soon die; David Livingstone and Fritzsche report that some nations in Central Africa kill children whose upper incisors protrude before the lower.

In considering the morbid processes which have been said to originate in normal teething, it should not be forgotten that dentition is a physiological process. As a rule, the gums, even when tumefied, have a pale color and show no symptom of inflammation; and generally there is no fever which can be demonstrated by the thermometer. There is no stomatitis, and no thrush, both of which are pathological conditions, though there is a certain amount of itching, even pruritus of the gums and a condition of irritation. There is very frequently a vaso-motor disturbance in the shape of reddened cheeks, but even this must not be attributed exclusively to the reflex irritation of dentition, since there are a great many conditions in which the same symptom is presented—for instance, pulmonary congestion, pleurisy, pneumonia, meningeal irritation.

It is also true that, now and then, there are slight muscular twitchings, and when the child is half asleep, the eyes will roll. There is sleeplessness, but we must not forget that peripheral irritability increases from the fifth to the ninth month considerably, and that the inhibitory centers do not perform all their functions as in the adult; thus it is even possible that a convulsion will occur, though I have not seen such a one dependent upon difficult dentition during the last ten years.

It is also stated that there are eruptions dependent upon normal dentition—urticaria, lichen, eczema; but it is questionable whether these have anything to do with the momentarily flushed cheeks of which I have spoken. We must not forget that about the time the teeth make their appearance congestion of all the parts of the head occurs uniformly; it is the time at which not only the teeth will protrude, but when the brain and skull develop to greater degree than ever during human life. Thus it is, that most cases of eczema, urticaria, etc., must be explained by uniform congestion of the parts, and not influences dependent upon dentition alone.

It has also been stated particularly by Vogel, that there is, now and again, a purulent conjunctivitis on the same side on which the teeth are protruding, and, Strümpell suggests that this may be, perhaps, the result of irritation extending from the antrum of Highmore and the nasal cavities—an explanation which seems very much strained.

It is also stated that pulmonary catarrh, bronchitis, and broncho-pneumonia are very frequent during, and in consequence of, dentition. It has been said that the catarrh may be the result of the large amount of salivation running out of the mouth upon the chest in such children. With regard to inflammatory diseases of the chest we must not forget that there are several causes which, about the time of dentition, are met with very frequently. It is the time in which children are more exposed to atmospheric influences. It must be remembered that within the first year the mortality among infants is greatest from diseases of the organs of digestion; in the second year from diseases of the organs of respiration, undoubtedly in consequence of the fact that during that period infants are more exposed to atmospheric changes than in earlier life. This is one of the reasons why, at the time of dentition, not in consequence of dentition, pulmonary diseases are frequent.

Another cause is that rhachitis is certainly on the increase in our country. It has always been more frequent than reputed, especially the form which is unattended by any considerable amount of glandular swelling. Even glandular enlargement need not be visible about the throat, but is perceptible in the deep-seated cervical glands, and in the lymph bodies of the mediastinum. These swollen glands give rise to bronchial catarrh, frequently to acute and chronic broncho-pneumonia, and not infrequently to phthisis.

Another ailment which is frequently attributed to dentition is diarrhœa. But is it found in most children who are teething? Certainly not. The large majority of infants who are either at the breast or whose artificial food is well selected, do not suffer from diarrhœa, while going through the process of dentition.

The occurrence of diarrhœa has been attributed to several causes; even to swallowing a large amount of saliva and the oral secretions which appear in children of three or four months of age, and continue a number of months. Nobody has ever stated that the copious salivation of the fourth or fifth month gives rise to diarrhœa; still, when the infants are six or seven months old, the diarrhœa is said to be the result of the same salivation!

Others have said that reputed dental diarrhœa is due to nervous influence showing itself in reflex disturbance of the splanchnic nerves. But the explanation has not been given, still the presumption prevails that this diarrhœa must be of a neurotic character.

It has appeared to me that the fear lest dentition should produce diarrhœa is very much exaggerated. At all events, the popular belief that there is such a thing as dental diarrhœa, has given rise to the practice of not caring for such flux, and many an incurable enteritis, and consecutive lymphadenitis and atrophy, has been due to the very fact that diarrhœa has been neglected. In all such cases, no matter whether diarrhœa or bronchitis are present, it is wrong to fall back upon dentition as the cause of these affections without looking for the diagnosis of something more. A large number of diseases which have been attributed to dentition, owe this erroneous diagnosis to the fact that the diagnostic powers of the practitioner were limited like those of the public with which he had to deal. This much I may add, that the local treatment of swollen gums, which consists in lancing, has fortunately become less common and popular than it was in former times, and although I see a large number of infants in the course of a year, I can state that in not more than two cases during the last five years have I felt called upon to carry out such a procedure. In a few instances I have done so under the impression that it might do good, inasmuch as the diagnosis was not quite clear; but in most of these cases I found, two or three days later, pneumonia which was quite easily determined, but which did not before develop sufficiently to prevent my making a mistake.

Is there anything which has not been attributed to the injurious influence of the second dentition? There are many amongst the public this very day (perhaps also amongst the practitioners?) who would coincide with E. Smith (Lancet I, 1869, p. 23), who expresses the conviction that the copious secretion transmitted from the oral mucous membrane is a very serious matter. According to him, the children become pale, thin, restless, appetite irregular, either diminished or exorbitant, bladder incontinent, constipation alternates with diarrhœa, worms are more copiously raised in the intestinal mucus; thus matters get worse and worse, until the child dies of phthisis. I should say that "phthisis" *might* and *ought to have been diagnosticated before, and perhaps prevented, if the dentition-ridden medical man had known how far to look after chronic glandular swellings, or chronic pleurisy or pneumonia, as the possible cause of the fatal termination.*

Of 100 deaths occurring in New York city in the course of one year, 29.63 took place in the first; 10.3 in the second; 4.37 in the third; 2.40 in the fourth; 1.64 in the fifth; 3.20 in the sixth year. Thus in the first six years occur 51.28 per cent. of all the deaths. The whole period from the end of the sixth to the eleventh year gives only 1.50 per cent. of all the deaths.

Thus there is considerable resistance on the part of the child's organism, after it has been fully developed to its seventh and eighth years.

There are some other facts which prove that this time is rather immune than otherwise.

Growth is most rapid in the first few years of life, not only as regards the head, but also the rest of the body. The length of the new-born is 18 inches; that of the adult 66 inches. The increase in the first year is 10 inches; in the second 4 inches; in the third 4 inches; in the fourth 3 inches; in the fifth 3 inches; in the sixth 2 inches; in the seventh, eighth, ninth and tenth, each 1 inch. Thus there is retardation of growth after the completion of the seventh year.

The proportion of the upper part of the trunk, that is the chest, to the lower, in the newly-born, is as 1 to 2; in the adult as 1 to 1.618. This normal proportion is attained with the eighth year.

The lumbar portion grows principally until the ninth year; then again between the twelfth and fifteenth, about the time of puberty.

Between the seventh and ninth years there is retardatiun of the growth of the lower extremities, as also the trunk and the whole body.

In the newly born, the proportion of the upper part of the head, the skull, to the lower part is 1 to 1; in the adult 1 to 1.618. This stationary proportion is attained with the eighth year.

After all, then, this is the time of the second dentition. Where now are the dangers to life?

Still, though not a serious danger, a great and permanent inconvenience and injury may originate between the first and second dentitions. They may result from the fact that the wall between the cavity of a temporary tooth grows thinner and disappears very gradually by permanent evulsion of the temporary teeth, particularly the bicuspids. The permanent teeth are very easily injured inasmuch as they are imbedded between the roots of the temporary ones. The damage done by that condition is frequently greater than that resulting from retardation in falling out, on the part of the temporary. But in the latter case, also, the beauty, position, and number of the permanent teeth can be impaired. Thus at this early time the advice of a professional dentist is frequently required.

There is but one good cause for premature evulsion of the milk teeth—namely—general periostitis or ostitis of the maxilla produced by inflammation of the root of the tooth.

It would be a mistake, however, to believe that we are more mediæval than other nations. The measures for relieving the dangers from the cruel attacks by the ambushing teeth, upon the unsophisticated baby, prove better than anything else how the maternal (and professional?) minds have been impressed by awe-stricken faith down to the second half of the enlightened nineteenth century. According to H. H. Ploss,* in different parts of Germany, Austria and Switzerland they resort to the following measures: A trowser button and the dried umbilical cord are kept under the pillow. The tooth of a colt a twelvemonth old is worn around the neck at the time of the increasing moon. The paw of a mole—bitten off—is sewed in and worn around the neck.

---

* Das Kind in Brauch und Sitte der Völker, 1876, II vol.

The baby to be licked by dogs. The head of a mouse to be used like the above mole paw. Every female visitor gives the baby a hard egg. The baby is carried to the butcher, who touches the gums with fresh calf's blood. The gums are touched with the tooth of a wolf, or the claw of a crab. The baby is supplied with three morsels from the first meal in the new residence after the wedding. Bread from the wedding feast of a newly married couple is in good repute. A mess of lindsprouts cut at twelve o'clock on Good Friday. A bone found by accident under the straw mattress. Mother, when first going to church after confinement, kneels on the right knee first. A man coming to visit, is silently given a coin, touches the gums of the baby three times and—goes to the tavern. So he does.

STOMACH.

The stomach and intestines participate in the general development of most organs. Between the embryonic state, and the individual after birth, or in advanced age, there is great histological difference. The former exhibits a rapid cell growth, the latter excels in tissue formation. In the stomach, after birth, the glands increase both in length, number and specific glandular structure; the mucous, submucous, and muscular layers while losing in cell growth, gain in fibre and density, and the lymph-vessels become narrower in width, and less in number, between the denser tissue. I have said in another part of this volume, that the stomach of the very young excels in absorption, that of the advanced child, in digestion. The same process is seen in the duodenum (and in the lower parts of the intestine), inasmuch as the villi and duplicatures increase in number and size and thus add to the surface, and the glands—mainly Brunner's—grow perceptibly.

The stomach and intestine in the fœtus and

newly born contain mucus and meconium, never air: no food having been given, the stomach contains air after half an hour. The intestinal tract then fills up gradually within the next twenty-four hours. Air cannot gain entrance except by swallowing, and even normal air produces fermentation and putrefaction; the air of a room or bed with puerperal fever may produce septic processes in the newly born, though no other inlet of poison existed in an individual case.

The capacity of the stomach is variable, and changes rapidly; from 35 to 43 cubic centimetres it increases to 153 or 160 within a fortnight, and 740 after two years.

The fundus of the stomach of the newly born is only slightly developed, and remains so until about the tenth month; during all this time it resembles greatly the stomach of the Carnivora. The empty stomach, in its position in the abdominal cavity is not flattened, as has been assumed. According to Fleischmann, the cardia, in the infant, is situated in front of the tenth dorsal vertebra, a little to its left; a pin run through the left margin of the cardia strikes the sixth left costal cartilage in the mammary line. When the stomach is expanded the cardia is not pushed sideways. The pylorus is fastened similarly, but not to the same extent. When the stomach is contracted or collapsed, the pylorus is always the lowest point, and found in the median line between the ensiform process and the umbilicus; in the majority of cases, the pylorus is the lowest point, even when the stomach is expanded. The pyloric portion of the infant's stomach does not extend beyond the median line of the body, while in the adult one-sixth part of the stomach extends beyond this point. Besides there is this difference: when the stomach of the adult is filled, cardia and pylorus are in almost a horizontal line, which is not so in the nursling; nor is the lower part of the stomach ever turned in such a way as to become the anterior aspect. The empty and contracted stomach of the infant is inaccessible to physical examination. The liver is very large, even in normal condition, and the left lobe is sometimes so massive as to entirely cover the stomach, and completely or nearly reach the spleen. Thus, whenever the stomach is empty, the tympanitic note heard over it belongs more to the colon and the rest of the intestine than to the stomach itself. Even when the stomach is filled, there is only a small triangular space, between the liver and the spleen, accessible to direct percussion; and sometimes the stomach is entirely covered by the colon; and whenever the liver is turned upward and forward, which has been said to be possible by filling the stomach, that change in position is due more to the colon than to the stomach.

GASTRIC DIGESTION.

Very young infants do not masticate; the secretion of saliva in the very first days is comparatively small, and its diastatic effect not instantaneous. Thus the nature of the food is, as it were, predestined, and its digestion left mostly to the stomach; hence it is that when nothing but good woman's milk is given, peptones are seldom or never found in the intestinal canal, or their remnants in the stomach. The cardiac and pyloric portions contribute but very little to digestion, and in the very young the peptic glands resemble in great degree the muciparous. Still, it is certain that the stomach of the newly born has been sufficiently prepared for gastric digestion; pepsin is present in the third or the beginning of the fourth month of utero-gestation; there is a good deal of it at the end of the fourth month: it was found a little later by Koelliker, according to whom the development of the gastric glands takes place in the fourth month only, while gastric acid is met with at a later period.

At that period of life the gastric muscles are feeble, with the exception of those layers which, beginning with the œsophagus, are distributed along the smaller curvature; the transverse fibres are thin, and the external longitudinal fibres are not perceptible on the pylorus. The pyloric valve is only slightly developed, and the part next to it is short and cylindrical. The fundus, but little developed, occupies an almost vertical position, and approximates that which is observed in the embryo or in the Carnivora. This cylindrical shape of the stomach, which, moreover, is imbedded between the large liver, the abdominal wall, and the almost perpendicular diaphragm, is the cause of the facility with which babies vomit. They as Schiff says of dogs, "do not enter into a long discussion with undigested food." This vertical position changes but gradually, but more readily only after large quantities of vegetable food have been partaken of, in the same way as the fundus of the stomach of dogs and cats is changed by mixed feeding.

There is no reason to believe that there is any material difference between the secretions and digestive processes in infantile and adult stomachs. The questions whether pepsin and gastric juice originate in the same glands, whether gastric juice is secreted in the same glands but in other cells, or whether hydrochloric acid is the result of secretion from muciparous glands acting on chloride of sodium, are physiologically the same in children as in adults.

Casein is coagulated by pepsin, to be dissolved again after a little time. This process takes place, even when the reaction is alkaline, with this exception that when it is alkaline the temperature required for coagulation is higher. The presence of acid appears to *aid* the formation of cheese (but it is not absolutely necessary). There is more pepsin about the fundus and pylorus, than in other parts of the stomach. Peptic coagulation, however, and coagulation by acids, act unequally; the latter takes place in finer flakes, the former consists simply in milky discoloration as long as the casein is unmixed. But casein which contains the phosphate of calcium coagulates into a thickish mass that contains a great deal of calcium and phosphoric acid. This last point is of great importance, for, in the hygiene of infants, an exaggerated importance is attributed to the copious or superabundant administration of calcium, a practice that is mostly useless, or dangerous.

There is another important point concerning the digestive powers of the infant, viz., the quantity of water in the food. This is especially important as to the secretion of, and the effect to be produced by, pepsin. The secretion of pepsin depends, to a great extent, on the nature of the food, and it is considerably increased by the ingestion of beef soup or solutions of sugar, or solutions of digested meat, and particularly of dextrine. Thus is explained the beneficial effect of soup taken before meals, inasmuch as it is quickly absorbed and increases the facility with which the pepsin required for the coming meal is secreted.

As soon as milk arrives in the stomach, water and some of the salts are absorbed, and pepsin is secreted. Butter is not here digested but undergoes its change under the influence of the bile and pancreatic juice, which, however, are only slightly efficient at the earliest period. Casein remains in the stomach, where it is under the influence of the digestive fluids which require large quantities of water; it is an old observation that water facilitates the digestion of caseine. Not only is the secretion of pepsin increased, but the formation of hydrochloric acid is improved by water. Consequently any cause which renders the food more concentrated than normal, disturbs digestion. Condensed milk requires considerable dilution. Increased frequency of sucking, which changes the milk, renders it more indigestible; also hot weather, fever, menstruation, and pregnancy of nursing women, produce the same effect, and require the copious use of water.

The formation of acid in the stomach depends mainly on the introduction of solid substances, while the formation of pepsin depends chiefly on liquids. When adults do not bear fluids it is because the proper proportion of acid and water, 4 to 1,000, has not been obtained In dyspeptic disorders of this kind, the introduction of a small quantity of hydrochloric acid, largely diluted, or, still better, the addition of more chloride of sodium to the food, is indicated.

In children this disproportion does not occur so frequently as in adults, as the former have a natural tendency to the formation of acid. Milk sugar changes into lactic acid, which is the *essential requirement for the first stage of digestion;* but water is often wanting. Small children are not, as a rule, supplied with water, besides, in the first months no fluid is secreted in the oral cavity which, when swallowed, might have a local effect in the stomach; thus it is easy to understand that infants usually have *too little* water with their food rather than too much. This is a very important reason why the food of infants should be given in fair dilution; which dilution may be even greater than the usual rules allow.

GASTRO INTESTINAL DISEASES—GENERAL REMARKS.

There is no class of diseases more frequent in the infant than those of the gastro-intestinal tract. They are somewhat more rare in advanced childhood, but originate on very slight provocations. As a rule, they are produced by errors in feeding, either as regards the quality or quantity of food; most infants and children eat too much and too often. The food should be chiefly of an albuminoid character, but this is very apt to undergo fermentation and putrefaction; thus it is that functional disorders very easily obtain. Such disorders of function will now and then run their course without producing many pathological changes; thus, dyspepsia, diarrhœa, colic, are names not always connected with a great many or extensive anatomical lesions. Still, we cannot imagine a disorder of function without a disordered organ, and it is always necessary to endeavor to find the connection between the clinical diagnosis and corresponding anatomical changes.

Disorders of the stomach and intestines are frequently found together, for diseases of the stomach will extend downward, and diseases of the intestine upward; therefore, the pathological and anatomical separation of the two is, in many instances, very difficult indeed. Diseases of the mucous membranes are apt to extend over large surfaces, besides, the lymphatic apparatus invariably participates. A catarrh is very liable to be transformed into an inflammation, the two are very often combined, and thus an acute process may become chronic, and chronic processes are apt to undergo acute relapses. All of these are conditions and changes that will occur very frequently.

In most of the disorders of the stomach and intestines which we are called upon to treat, there is only tumefaction of the mucous membrane and the glands, even in those cases in which the symptoms are very serious and the consequences dangerous. Thus the diagnosis between chronic and subacute catarrh, subacute and acute catarrh, acute catarrh and inflammation, cannot always be made with absolute certainty; nor is it necessary, so long as it is remembered that these several conditions will merge into each other both clinically and anatomically. There are unmixed and uncomplicated cases, but they are rare indeed.

DYSPEPSIA.

One of the disorders of function, dependent sometimes upon slight changes in the gastric mucous membrane, is called dyspepsia, a term intended to convey partial or complete loss of appetite with slow or absent digestion. In regard to this, however, it is not well to rely too implicitly upon the tales of mothers or nurses. Older children will complain of præcordial heaviness, and will suffer, as do infants, from eructations, which when they result from swallowing air, are absolutely odorless, and, when they consist of actual gastric gases, have a very faint odor. A sensation of oppression and of pain in the forehead is complained of by older children, while younger ones are apt to vomit.

The causes of dyspepsia must be sought for either in anatomical changes in the stomach, which can rarely be proven; or in quantitative or qualitative changes in the secretion, which are more frequent; or in a changed nerve influence, as for instance, in fever; or in an abnormal condition of the food, which is the most frequent cause.

The treatment of this disorder consists chiefly in abstinence or in the use of the most simple diet. On this point I shall have more to say elsewhere.

In infants who have been fed artificially, the gastric secretion is apt to be more than usually acid. In those cases alkalies should be given at once. The treatment of adults in such cases has, in late years, consisted in washing out the stomach. To what extent, in the future, this practice can be made useful in children and infants, remains to be seen. There are some who speak very favorably on its indications and successes.

NERVOUS DYSPEPSIA.

The digestive organs are easily influenced by the nervous system. Emotions will give rise to nausea, vomiting, pain, diarrhœa, particularly in so-called nervous people, or when hysteria or neurasthænia are fully developed. When such influences act frequently, and are combined with hypochondriacal impressions, we speak of nervous dyspepsia, such as is met in men, women, and—less frequently—in children.

The symptoms vary. There is either loss of appetite or craving for food—the stomach is very sensitive to food sometimes, but other times it is not sensitive at all, and even errors in diet do not result badly. Pressure on the stomach is sometimes painful, and sometimes it gives relief. Vomiting will often occur without any connection with feeding; is sometimes acid, and sometimes the eructations have neither taste nor odor. There is a good deal of audible peristaltic rumbling, and the abdomen is generally distended. The movements of the bowels are irregular. There is usually a number of other symptoms, such as irritability, hypochondriacal sensations, vertigo, headache, and paræsthetic disorders such as coldness and numbness, and prickling sensations. General nutrition is sometimes good, but in most cases, the patients are pale and anæmic. Objective examination of the parts is, as a rule, negative.

Hysteria, neurasthænia, and nervous disorders, complicated mostly with anæmia, are by no means so rare in children as has been supposed; all the symptoms reported of the adult woman may be present in the young, both male and female. Such conditions of nervous dyspepsia in the young, particularly in girls between six and twelve years of age, I have seen frequently; they will be observed by anyone who will watch for them. The causes in general are premature mental exertion, hereditary predisposition, hysteria, congenital or acquired anæmia.

Therapeutics must be simple, yet the effect is not very encouraging. Food must be digestible and copious. Purgatives should never be given; enemata must take their place, if required. Bitter tonics, mild preparations of iron, country and sea air, cold bathing or sponge baths, electricity, one large electrode being applied to the stomach and another to the spinal column, are indicated.

POLYPHAGIA—BULIMIA.

In some forms of gastro-intestinal disorders there is no loss of appetite. On the contrary, there are even babes and children who suffer, to an exhorbitant degree, from what appears to be appetite or hunger; there is a constant craving for food. Sometimes it is the result of bad habits, in which case the infant or child will look for sweet articles of food mainly. In other cases, however, there is a substantial anatomical reason for the affection, which is to be looked for more especially when the young patient does not increase in weight, or does not hold its own, while the symptoms still continue.

A frequent cause of this disorder is the presence of an excessive number of intestinal worms; in many we have to deal with hypertrophy of the mesenteric glands, either of "scrofulous" origin or, which is more frequent, due to neglect of chronic diarrhœa; and in a few cases the symptom is due to chronic disorders of the brain.

VOMITING.

Vomiting, one of the symptoms which accompany dyspepsia, in the infant, is almost a normal occurrence. The infantile stomach is vertical and cylindrical, and the fundus is but little developed; thus whenever there is a tendency to empty the stomach, the anti-peristaltic motions do not press against the fundus, but directly upwards; there is, therefore, less real vomiting, than mere overflow of contents, which takes place so easily that the babe is not disturbed by it.

The treatment of such cases, if treatment be required at all, consist chiefly in the application of dietetic rules. The infant should have less food, and at longer intervals; should not be carried about immediately after meals; ought not to be shaken or jolted; nor carried face downwards.

The overflow takes place, as a rule, immediately after the baby has been nursed. If so, the milk is still fluid. If vomiting occurs a little later, the milk will be coagulated; if then the milk be not coagulated, the stomach is not in a normal condition. In these cases, and particularly when the baby lives on artificial food, there is uneasiness and pain associated with the vomiting, and an acid mucus is expelled with contents of the stomach. These are the cases in which anti-fermentatives, such as nitrate of silver, bismuth, resorcin, etc., are indicated; sometimes antacids alone will suffice.

In a number of cases vomiting is preceded by more serious symptoms. Now and then paleness and a condition of collapse are present, which last until the act of vomiting is completed. In order to terminate the disagreeable symptoms it may be necessary to accelerate the vomiting, which should not be done by emetics, however, unless their need becomes absolute, since, as a rule, friction of the epigastrium and tickling the throat is sufficient to produce the desired effect.

Mental emotions, such as fear and fright, may produce vomiting in older children. A very frequent cause of vomiting is the presence of ascarides, which have left their habitat in the intestine, and crawled upward into the stomach, and now and again, even without exciting vomiting, make their appearance in the mouth, and, in rare cases, pass through the nose, and in still rarer instances, through the ears.

Vomiting is also a symptom in the incipient stage of acute diseases, as, for example, intermittent fever and pneumonia; particularly also of poisoning, either by poisons introduced in the stomach, or such as circulate in the blood, including those of acute febrile eruptions, especially scarlatina. Finally, diseases of the brain with irritation of the pneumogastric exhibit, among the very first symptoms, vomiting, which is especially liable to take place when the children change their position, or are raised suddenly.

ACUTE GASTRIC CATARRH.

The feeble, the anæmic, the convalescent, and the feverish are predisposed to this affection; but it may occur in the previously healthy as well. In all such children the production of gastric acid is diminished, and thereby digestion is impaired; besides, in all, muscular power is reduced.

Cold or hot ingesta, too large quantities of food, acids, spices, irritant medicines, alcoholic drinks, fat meat, cake, decomposed food with its ferments, may one and all be the cause of acute gastric catarrh. That dentition, *as such*, is *not* a cause, has been mentioned already. Exposure to sudden changes of temperature is certainly apt to produce it.

Opportunities to study the pathological anatomy are not frequent, for uncomplicated cases are seldom the subjects of post-mortem examinations. In all, however, the epithelium is swollen and changed in its composition; the mucous membrane is injected, pink, thickened, and folded, and its surface is gray with transformed epithelium and mucus.

Symptoms.—Not infrequently there is fever, with a temperature of 104° F., or more; and when the child is young, the temperature high, the local irritation considerable, convulsions are frequent. There may be vomiting when the material expelled is very copious, flowing through the mouth and nose at the same time, is acid, and contains food, mucus, and later, bile. Usually there is a great deal of thirst; the patient wants water all the time, and swallows it greedily. The tongue is seldom much changed in the beginning. There is pain in the stomach and precordial region; older children complain of a sensation of fullness. In the later stages the mouth is red and dry, the tongue furred, and an acid odor to the breath, but without the fœtor that obtains in adults. There is no appetite, but some stupor, and sometimes delirium. Litten found the peculiar odor of fruit, also aceton, and thus compares the nervous symptoms of severe acute gastric catarrh with those of the final termination of diabetes. The urine has a somewhat high color, and is diminished in quantity; generally also there is constipation. Respiration is accelerated in proportion to the temperature only. The inflammatory or catarrhal condition may extend downward to the duodenum, and give rise to jaundice; sometimes so frequently that epidemics of *gastro intestinal icterus* have been observed (Rehn, Flesch). In this condition the child remains for one or two days, then it begins to improve, the mouth being the last to return to the normal condition; at this time labial herpes may appear. Diarrhœa will set in now and then. Appetite will return slowly, but not infrequently increase into hunger, when great precaution is demanded with reference to diet. Occasionally, too, when no care has been taken, chronic catarrh will result from an acute attack.

The diagnosis of acute gastric catarrh may be easy enough when the history of the case is simple, but the reverse is true when there is no history and the symptoms severe.

Convulsions may occur from primary meningitis, mainly in the purulent or fibrino-purulent forms, which run an acute course with few or no premonitory symptoms; but in primary meningitis, other symptoms, for instance those belonging to the pupils and to the central nerves in general, will show themselves; besides the convulsions in acute gastritis are apt to cease after a while and not return, except in those cases in which the first convulsion produces permanent central symptoms by an effusion or extravasation.

Convulsions are also possible in pneumonia, which now and then begins with one or more. When the fever remains high for several days, it speaks more for pneumonia than for gastric catarrh, although no physical signs of pneumonia may be found. Indeed, in many cases pneumonia, when it begins centrally, cannot be detected by auscultation and percussion until a number of days have passed; but the proportion between the pulse and respiration is very soon changed, so that the normal ratio of ten respirations to thirty-seven or thirty-eight heart-beats is disturbed. When there is such a disturbance, and there is a greater number of respirations in proportion, the respiratory organs must be suspected. Moaning pneumonic respiration may be met with in colic and also in tenesmus, but in pneumonia it is apt to be constant, and in colic and tenesmus will also be attended by other diagnostic symptoms.

Typhoid fever may appear, particularly in older children, quite suddenly, with high temperature; this occurrence is exceptional, but is occasionally met with. The slowly increasing scale of temperature, of which so much can be read, is not so often seen in practice. When I consider the course of typhoid fever in the young and in the adult as I see cases in New York, and hear of them in other States, it appears to me that the books have copied from each other, and not from observations of Nature. Typhoid fever will not be attended by the labial herpes seen in acute gastric catarrh.

The prognosis is favorable, although in the very beginning, during the convulsions, central hæmorrhages are among the possibilities, and relapses are frequent.

Treatment.—Remove the injurious substances from the stomach. If vomiting has not occurred or has not been sufficiently profuse, it should be provoked by tickling the fauces, and friction of the præcordial region. These, as a rule, will suffice, but if they do not, quantities of warm water or mustard water will answer a good purpose. Either ipecac or turpeth mineral, is a good and efficient remedy, also apomorphia, either internally or, if necessary, by subcutaneous injection; the syrup of ipecac is a doubtful and unreliable preparation. Purgatives should not be given in the beginning; large enemata will act more favorably—warm water, or warm water with anti-spasmodics such as assafœtida, or local stimulants such as turpentine. After a day or two, calomel, two to six grains, according to age, will answer well.

Fever, unless it be high, requires no special treatment; in urgent cases only ought antipyrin to be given. As a rule, cold applications to the head act well when there is a tendency to convulsions; cold applications to the heart reduce the temperature of the whole body. A warm bath will frequently do good, but I *do not advise bathing or handling the child much while the convulsion is on.* When thirst is great, small quantities of ice-water should be given often, or Seltzer, Vichy, or Apollinaris; also water to which dilute muriatic acid has been added in the proportion of one, to three to ten thousand.

Solid food must not be given. When there is a great deal of mucus, milk must be given very much diluted, or peptonized. When the tendency to vomit is great, food and drink must be given in teaspoonful doses, and where the sensitiveness of the stomach is very marked, mucilaginous and farinaceous foods only will answer, together with small doses of bismuth repeated every one or two hours. Where acid is predominant, calcined magnesia will answer best, given in small frequently repeated doses; also bicarbonate of sodium, although the latter does not work so well because of carbonic acid set free in the stomach.

In cases of constipation magnesia alone will usually suffice, or tablespoon doses of Carlsbad or Congress water, may be given every one or two hours. The German aqueous tincture of rhubarb in doses of 10 to 30 minims every two or four hours, also answers well.

CORROSIVE GASTRITIS.

The deglutition of strong acids or alkalies is apt to result fatally in a very short time. When a fatal result does not occur, there is, besides the local affection, an inflammatory reaction around its margin; afterwards, ulceration and cicatrization. When the effect is local, it is generally found in the large curvature opposite the cardia. In mild cases there is hemorrhagic erosion, punctiform or longitudinal. The surface is denuded of its epithelium or there may be deeper loss of substance. The lesion can be easily distinguished from tubercular ulceration.

In mild cases the symptoms are those of acute or chronic gastric catarrh. In severe cases there is persistent vomiting of mucus of a bloody character. There is pain and fever.

The local effect produced by nitrate of silver differs from the above. When nitrate of silver in substance touches the mucous membrane, its effect is quite local. It will stick to the place it first touches. It forms an albuminate on the surrounding mucous membrane, and is very apt to be buried in its own albuminate, thereby becoming partly innocuous for some time. Thus I found it to be in the case of a little child that had swallowed a stick of nitrate of silver which was applied for the purpose of cauterizing the throat in a case of presumed diphtheria. It was, moreover, a case in which no chloride of sodium had been given as an antidote.

In the cases of poisoning by lye or acids, these should be neutralized, in the first instance by diluted acid (lemon juice, vinegar,) and in the second by dilute alkaline solutions (lime, magnesia, soap).

When plenty of neutralizing fluids have been taken, the stomach pump may be used, and the organ washed out. But the greatest precaution must be taken in introducing the tube. I once saw an œsophagus that had been perforated by a bougie in the hands of a careful man; and the child died. In the right pleural cavity there were twelve ounces of milk and whiskey.

The only thing to be done for some time in this class of cases is to administer ice, milk and opium Later, nitrate of silver greatly diluted, bismuth, and weak solutions of chloral hydrate. Baginsky recommends iodoform.

DIPHTHERITIC GASTRITIS.

This occurs rarely, and then only as a part of general diphtheria. The deposits on the mucous membrane of the stomach are small, detached, and macerate very easily. In these cases there is bloody vomiting and pain, together with the other symptoms of diphtheria. Large masses of diphtheritic exudation are found less in diphtheria proper than in variola or typhoid fever of severe character.

A five year old child seen by me twenty-five years ago, had the œsophagus completely filled, and the stomach lined, with a thick croupous membrane like that filling the bronchi in cases of fibrinous bronchitis. There was no bloody vomiting and the death of the child presented more of the general symptoms of typhoid fever than of the local affection.

SUPPURATIVE GASTRITIS.

This consists in the formation of small or large abscesses in the wall of the stomach. The location of these abscesses is in the submucous tissue, but they may extend through the muscular layer to the serous membrane, thus giving rise to perforation. Fortutunately such are very rare indeed; the only case I have ever seen in an infant, was the result of pyæmia.

CHRONIC GASTRIC CATARRH.

Hereditary predisposition to this affection has been observed in whole families, and may be the result of acute gastric catarrh, without reference to the cause of that disease. It follows the persistent ingestion of injurious food, and erroneous diet in general; is very often, in older children, complicated with intellectual over-exertion, and insufficient attention to the skin; and results, not infrequently, from general disturbances of the circulation, particularly those which depend upon chronic adhesive pleurisy and pneumonia, chronic cardiac disorders, and disturbances of the portal vein brought about by too much schooling and during slow convalescence attended by constipation. It is the result, also, of local irritation; the influence of bad air, particularly when heated; impaired innervation; masturbation in younger and older children; impairment of the general health by rhachitis, which is so apt to influence muscular action; and tuberculosis, caries, and "scrofula."

The disturbances of circulation alluded to, diminish the secretion of gastric juice, particularly of muriatic acid which is needed more than the pepsin. Thus the products of fermentation are formed, such as lactic, butyric, and acetic acid, which will increase the chances of the development of catarrh. If there be a large quantity of mucus, it covers the ingesta, thus preventing the gastric juice from exerting its normal influence on the food; besides, mucus has an alkaline reaction and diminishes the effect of the gastric secretion.

Peristalsis of the stomach is impaired by catarrh,

inasmuch as the muscular layer becomes œdematous; in that condition the stomach is more apt to be expanded and undergo dilatation by the introduction of large quantities of food.

Disturbances of circulation, by preventing normal absorption will, moreover, prevent the absorption of peptones present in the stomach; peptones formed out of food must be absorbed immediately, as otherwise they are known to disturb the peptonization of the albuminates. In all those cases in which peristalsis is deficient the peptones are apt to remain behind.

The anatomical changes in chronic gastric catarrh are not always considerable. There is a large quantity of viscid and grayish mucous mixed with much epithelium upon the mucous membrane, which membrane is mostly red, or red in patches, sometimes gray by deposit of pigment, and in small spots ecchymoses over which the epithelium is very apt to be absent; most of the changes are found in the pyloric portion. After a while the surface becomes smooth, and is apt to get atrophied; also in protracted cases the glands, which were first swollen, undergo atrophy, while the cellular tissue between them may be thickened, the whole mucous membrane hyperplastic and folded. The submucous tissue may also be thickened and show a large increase in the number of cells, and the same condition may be present in the muscular layers. In some cases of chronic gastric catarrh the stomach is somewhat dilated.

Symptoms.—The function of the stomach is diminished, and there is a constant tendency to acute exacerbations; children suffer from loss of appetite, alternating sometimes with crying and hunger; the last, however, is very easily satisfied. There is great thirst, and a longing for pungent, highly-seasoned food; occasionally there is vomiting, sometimes immediately after meals, sometimes later. There is considerable nausea, acid taste and eructations, depending upon the prevalence of organic acids, but not hydrochloric acid. There is seldom any blood in what is vomited, but abundance of fermentation products and sarcinæ.

The stomach is inflated and tense; there is not exactly pain, but a sensation of fullness and discomfort. The tongue is grayish with reddened edges, and now and then there are superficial erosions in the mouth, and considerable salivation. Constipation is not infrequent; sometimes diarrhœa alternates with constipation. The skin is pale and flabby; emaciation takes place frequently; sometimes there is jaundice. The urine contains a large amount of phosphates and is slightly acid. When the disorder is of long standing, nutrition is so much impaired that chronic cutaneous eruptions are either apt to result or, if they existed before, will not get well. Acute exacerbations are not infrequent. Improvement takes place but gradually, and there is a tendency to relapses, the more so as the albuminates of infants are apt to become decomposed.

The *diagnosis* of this affection is easy enough when it sets in after an acute attack of gastric catarrh, or when the cause is well known; otherwise it may be very difficult. Tubercular meningitis is sometimes mistaken for gastric catarrh, inasmuch as vomiting is very often the very first symptom of that disease to which attention is directed; but at the time, when the vomiting is observed in tubercular meningitis, there are other symptoms in almost every case:—the pulse is irregular and slow; not infrequently there is severe headache, either constant or paroxysmal; and for weeks previous to the vomiting there are psychical changes, such as peevishness, fretfulness, unwonted quiet, etc. It is true that mistakes are possible on the part of the physician who does not observe the child more than a few minutes; but this being so, he should always remember that a mistake can be avoided by close questioning.

The result of chronic gastric catarrh has now and then been mistaken for the mal-nutrition and emaciation attending chronic processes in the peritoneum and lungs, consequently in any doubtful case, physical examination should not be omitted.

Typhoid fever may be mistaken for chronic gastric catarrh, and *vice versa*. Both of these diseases may run their full course with very little fever.

Chronic nephritis may be suspected in some cases, but an examination of the urine will clear up the diagnosis.

The diagnosis from other affections, such as ulcers and dilatation of the stomach, and nervous dyspepsia, will be discussed elsewhere. Finally, chronic gastric catarrh should never be diagnosed except when other diseases can be safely excluded.

Prognosis is fair, provided great attention is paid for years to hygiene and diet, and acute exacerbations are treated carefully with appropriate medicines.

Treatment.—Errors in diet must be avoided, and the quality and quantity of food controlled. Most children eat too much. But little solid food is to be given; no sweets, no fats. The morsels must be small and mastication slow. An acute catarrh must be healed. Masturbation must be prevented. Sedentary life must be forbidden. School hours must be limited, and interrupted by gentle exercise, further, they must be regulated by the meals, and not *vice versa*.

When there is a chronic disorder of the heart or lungs or liver, it must be attended to. Many a gastric catarrh in the child, or in the adult, will be improved by digitalis. When there is constipation, enemata will act well. Rhubarb and magnesia, rhubarb and bicarbonate of sodium, or the aqueous tincture of rhubarb in small doses frequently repeated, will answer quite well. When there is a great deal of mucus, dilute muriatic acid mixed with plenty of water and pepsin, will serve a good purpose. When the tongue is thickly coated and there are eructations, the chloride of ammonium with the aqueous tincture of rhubarb will prove beneficial. The pain and tendency to vomit require bismuth; in older children bitter waters, Carlsbad, or Congress. The practice of washing out the stomach may come in very well in older children; in younger children it has been resorted to by Demme, when there was a great deal of fermentation, and digestion was very slow. Massage though ever so gentle must not be attempted while the stomach is full.

Meals ought to be scanty but more frequent; toasted bread and stale bread, answer very well. Milk must be diluted, either on the plan of Rudisch or by farinaceous decoctions, or it may be peptonized. Alcohol, and too many carbohydrates, must not be given, and the food must be neither too hot nor too cold. When digestion is slow the addition of chloride of sodium, or sodium bicarbonate, and effervescent drinks in general, will often be efficient, because they stimulate the secretion of gastric juice where there is an excess of fermentation; creasote in doses of $\frac{1}{4}$ of a minim to 1 minim, three or four times a day in plenty of water, will serve a good purpose; also a few grains of salicylic acid through the day, largely diluted with water; benzine has been given, also, in doses of a few drops in water or milk. A chronic gastric catarrh ought to be treated for a long time with bismuth, nitrate of silver, or sulphate of zinc.

Where there is much vomiting, the occasional

administration of small pieces of ice will answer. Also in obstinate cases, very small doses of opium, chloral, or Fowler's solution (¼ m. to ½ m., three to eight times a day), act very well when everything else appears to fail.

Constipation must not be treated with medicines, for sometimes there is *apparent constipation* depending upon the diminished quantity of food given or entering the intestinal tract.

Anæmia must not always be treated by the use of iron; it is badly tolerated in most gastric disorders. But bitter tonics, such as nux vomica or condurango, are well tolerated, and beneficial. Mountain or sea air will improve digestion and sanguification.

DILATATION OF THE STOMACH.

This does not occur so frequently in infants and children as in adults. The causes in the former are:

First. Amylaceous diet, and superabundance of food in general, which result in over-exertion and expansion, with voracious appetite.

Second. The voracity of rhachitic children, with their feeble muscular tissue, imperfect digestion, and consecutive gaseous inflation.

Third. Catarrhal inflammation resulting in œdematous softening, particularly in chronic conditions of the same kind, and an increased expansion with diminished absorption.

Fourth. General muscular debility, as in anæmia and convalescence.

Fifth. Congenital imperfection, sometimes even partial absence of muscular development.

Sixth. Peritoneal adhesions of the walls of the stomach, resulting not only in dilatation, but in deformities of different kinds. Such stomachs are more or less triangular, or quadrangular.

Symptoms.—These include all the symptoms of catarrh of the stomach. The epigastrium is distended but percussion does not yield very conclusive results, for the normal stomach of the infant and child is accessible with difficulty; thus when the stomach is dilated the patient ought to be examined both while lying and standing, and while the stomach is both empty and full.

The percussion sounds are apt to be rendered indistinct by the pulmonary sounds above, and the intestinal tympanites below. Sometimes in the half-filled stomach there are splashing noises on slight percussion. The large curvature is often visible, particularly on gentle friction. Very often there is craving hunger, and at the same time emaciation takes place, and there are eructations sometimes of acid, air, and liquid; also vomiting, frequently of large masses resulting from the meals of the day or two days previous. A brown-yellow diarrhœa alternates with constipation, the latter being due, now and then, to an insufficient amount of food entering the intestinal canal.

For diagnostic purposes bicarbonate of sodium

and tartaric acid have been introduced into the stomach separately, but closely together, to distend it so that it could be measured, but this method yields only uncertain results, inasmuch as the viscus will be dilated by the gas beyond its ordinary size. Whether the stomach sound will be of any service remains to be seen.

But little liquid is absorbed; thus the urine is not copious, and the dryness of the tissues in general results in spasm, particularly of the flexors of the arms, the calves, and the abdomen.

According to Ewald, there is a symptom which may be promising, based upon the changes which take place in salol, which decomposes in the small intestines only. As long as the stomach is normal, salicylic acid appears in the urine, as the result of the decomposition of salol, in from one-half to three-fourths of an hour after this drug has been taken; while in dilatation of the stomach from two to three hours are required. The gross appearance of the urine when salicylic acid is present is about the same as when it contains carbolic acid.

The *prognosis* is better in children than in adults. It is best when it depends upon those conditions which can be improved or cured. It is fair in cases of muscular weakness dependent upon rhachitis, and in catarrh caused by improper feeding.

Therapeutics.—The treatment is prominently that of chronic catarrh of the stomach. Anti-ferments must be given, such as bismuth, nitrate of silver, calomel, and resorcin. The quantity of food taken at one time should be small, but the meals should be frequent. Nothing should be given that is apt to ferment, like starch; large quantities of fluid should not be given, though milk may be allowed in small quantities. Diarrhœa may require tannin and other astringents. As it depends upon the condition of the stomach, most cases of consecutive diarrhœa will be best treated by attending to the digestion. Raw beef is amongst those articles of food which are most easily digested. Raw milk is not so easily digested as boiled, and peptonized milk or milk prepared according to the suggestion of Rudisch is preferable in many cases. A bandage should be worn about the abdomen. The Faradic and Galvanic currents can be used with advantage; according to Ewald, electricity and massage accelerate the passage of chyme into the intestine, but it seems to me that it is questionable whether digestion is thereby improved, for it may be that both of these applications result in premature opening of the pylorus, — before gastric digestion is finished. Nux vomica internally will improve the tonicity of the gastric muscle.

Kussmaul and Leube have introduced the mechanical treatment of the stomach by washing it out half an hour before the principal meal, either with water, or with a solution of bicarbonate of sodium in water, one or two per cent., or a one-per-cent. solution of salicylic acid, or a two-per-cent. solution of resorcin.

HÆMORRHAGE FROM THE STOMACH.

Blood expelled by vomiting is coagulated and more or less black. The general symptoms depend upon the quantity of blood lost. There is momentary paleness, and more or less consecutive anæmia. Now and then black blood, viscid and tarry, will be evacuated from the bowels.

The cause is very often local ulceration, sometimes, however, merely venous hyperæmia of the wall of the stomach; thus pulmonary and cardiac diseases predispose to extravasations into the stomach, as does also thrombosis of the portal vein, and cirrhosis of the liver, which occur occasionally even in childhood. Besides, gastric hæmorrhages are observed in acute atrophy of the liver, which, although rare, may occur, in connection with general hæmorrhagic diathesis, scurvy, morbus maculosus, and leucocythæmia.

Applications of ice to the stomach always do good, opium also is beneficial; but I have not observed that any preparation of iron has been of service, on the contrary, it has appeared to me that the coagula, which are formed in the stomach by ferruginous preparations, as the subsulphate or perchloride, act as irritants and produce new vomiting.

ULCER OF THE STOMACH AND DUODENUM.

The round perforating ulcer of the stomach is found more frequently in the new-born and quite young than in advanced childhood, though I have met with it in children from seven to twelve years of age, several times. It is rare, however, in children under ten years old, but about that age it occurs in girls who show distinct signs of chlorosis. Brinton saw two patients under ten years of age, in 226 cases. Biedert saw a girl of twelve years who died after two days of exhaustion following gastric hæmorrhage. A number of cases have been observed in tuberculosis and after measles and scarlatina.

The causes of gastric and duodenal ulcers are numerous. Some enumerate among these arterial anæmia, others venous hyperæmia or stasis in the hepatic vessels, or circumscribed hæmorrhages into the tissue; others again assign thrombosis as the cause; others, emboli or lessened alkalinity of the blood, or hæmoglobinuria—in the last mentioned affection, as also in the later stages of diabetes, the blood is slightly acid. In this way hæmorrhages and ulcers go hand in hand frequently, and are often the cause of each other, both in the adult and in the young.

In the newly born, hæmorrhages from the digestive tract are known by the name of *melæna*. They result in the vomiting of black blood, also rectal evacuations of the same nature and color, and tar-like consistency and appearance. In the majority of cases melæna occurs from localized and circumscribed ulcers, or from the rupture of blood-vessels between the first and third day of life, rarely after the first week; it is caused also by the sudden changes in the circulation occurring at and after birth; at that time, and from the same causes, may occur dozens and hundreds of small hæmorrhages of the pericardium, pleura, and also in the mucous membrane of the stomach and duodenum.

Hyperæmia results in hæmorrhage very much more frequently in the newly born and infants, because of the thinness and permeability of the walls of the blood-vessels. Thus the frequency of meningeal hæmorrhages in the very young finds its ready explanation. Hyperæmia and hæmorrhage with their dangerous consequences are met with in vigorous as well as in feeble infants; those, however, who are born asphyxiated, are more apt to suffer from them and their posssible results. The principal consequence is ulceration of the mucous membrane into which copious extravasation has taken place. Regular hæmorrhagic infarctions are found in the stomach and duodenum of the new-born, produced by emboli derived from the thrombi of the umbilical vein and duct of Botallus. They cause both hæmorrhages and ulcers. Another cause of these hæmorrhages, which are complicated with, or are the causes of, ulcers, is acute fatty degeneration in the fœtus, a process found mostly in the epithelium and the endothelium of all the organs, and also in their tissues, and that renders the blood-vessels and surrounding tissues more fragile; it is a frequent cause of gastric, duodenal, and general bleeding. Many of the cases of so-called hæmophilia in the newly-born, particularly those occurring in families where there is no history of hæmophilia, are, indeed, cases of fatty degeneration. It is the same process that is seen at every age as the result of poisoning by phosphorus, arsenic, antimony, acids, or intense heat.

When ulcer of the stomach or duodenum has been produced by any of the above mentioned causes, or when it is formed either as the result of old chronic catarrh, or by sudden interruption of the circulation in a circumscribed part of the mucous membrane, thus diminishing or destroying the normal alkalinity of the tissues, the dangerous features are various.

The constant presence of acid in the stomach and upper part of the duodenum will digest the parts which have been denuded or eaten, into, and are no longer in connection with the normal alkaline circulation, and no longer protected against the surface acids; these parts are in a condition similar to the dead gastric tissue in which gastromalacia is formed at autopsies.

Thus the first indication is to *keep the stomach and duodenum as alkaline as possible*, at all events between meals. The introduction of any food will give rise to the secretion of gastric juice, which contains first lactic, afterwards hydrochloric acid. It is secreted in quite large quantities, but normally spent on the digestion of food. Whatever there is, however, in the stomach, acids not required for the physiological process, particularly the acetic, the butyric, caprylic, or even an excess of lactic, all that must be neutralized. An occasional dose of an antacid is not sufficient for that purpose, but it must be given regularly; I generally give the doses at intervals of two hours. I also give a dose a few minutes before each meal to neutralize every abnormal acid. Of the antacids, I prefer calcined magnesium to the carbonate, as I do not wish the expulsion of free carbonic acid into the stomach; I use it frequently, but rarely for a child in larger doses than eight to ten or twelve grains daily. A small part of this, say one grain, is taken. every hour or two, before meals, mostly in water, which should not be too cold; hot water is even better. More than that quantity is seldom tolerated because of diarrhœa setting in, still its purgative effect is very welcome in patients suffering from constipation, who may take larger doses. When the above quantity does not suffice to neutralize the acids, or it is feared that more magnesium will cause dirrhœa, it may be combined with carbonate or phosphate of lime. Sodium bicarbonate does not take the place of the calcium and magnesium so readily, inasmuch as it also appears to promote the secretion of gastric juice. Thus, in most cases, I use magnesium or calcium with or without bismuth or with such adjuvants as may appear to be indicated for other reasons. This medicinal treatment must be continued for weeks or months; without it I do not see gastric or duodenal ulcers getting well. Carlsbad waters, and salines in general, owe their effects, partly to the neutralizing, and partly to the purgative influence they exert. The use of lime water is in part an illusion, if given for the purpose of neutralizing the acid. It is a failure because it contains only a single grain to nearly two fluid ounces of liquid; but when added to cow's milk it certainly makes it more digestible.

Another danger is the persistence of the function of the diseased organ. Both the stomach and the duodenum should be kept as idle as possible, and their labor should be made easy. Undigestible food must not be given, and solid food must not be allowed. Most cases, in older children, bear boiled milk, strained oatmeal, barley gruel, stale wheat bread, and a few also raw beef; some take nothing but boiled milk, or buttermilk, or koumyss. Many, particularly convalescents or adults, will tell you that they do not digest milk. That may be true, but then they gulp it down and it forms a caseous cake in the stomach that is not afterwards dissolved and digested. They must *boil their milk* in the morning, *heat it several times* during the day almost to boiling point, must add a small quantity of *table salt* to it, and in case the stomach is very acid, some bicarbonate of sodium, or calcium, or magnesium; neither must they drink the milk, but pour it upon a plate and sip it with a spoon. Thus prepared, they will digest it, particularly when it is not quite cold; in fact many require their meals warm or hot.

For the purpose of easier digestion, milk may be peptonized, or may be rendered more digestible by the process recommended by Dr. Rudisch, which consists in mixing one part of dilute muriatic acid with 250 parts of water and 500 parts of raw milk, and then boiling; in this mixture the muriatic acid is kept in the form of chlorides. It does not do so well in ulcers, however, as in acute or chronic gastric catarrh. Mixing milk, prepared as above, with some farinaceous decoction, will aid digestibility and add to its nutrient qualities. On such food, which must be adhered to strictly, the patients will do well and recover entirely. Many of my patients have lived on milk and butter-milk for months and got well. In the latter periods of the disease peptonized meats may be used.

With an alkaline condition of the surface and an innocuous diet, the ulcers have an opportunity to heal. Their recovery may be further aided by the administration of nitrate of silver, of which a child may take from $\frac{1}{30}$th to $\frac{1}{20}$th of a grain in a teaspoonful of distilled water four or five times a day, if possible on a fairly empty stomach; or, a smaller quantity may be given in a pill with or without a small dose of opium, say $\frac{1}{60}$th to $\frac{1}{15}$th of a grain in each pill. Sometimes I gave but a single dose at bed time, in addition to the alkaline treatment. Tincture of iodine, in half to one-drop doses for children, well diluted with water, has often been recommended, and its action is probably antifermentative.

When there is much pain and a great deal of acid or other secretion, opiates are indicated as above, mainly those which are very soluble. Chloral is tolerated badly. Gerhardt prescribes, particularly in cases where pain is a prominent symptom, from three to four drops of the liquor ferri perchloridi in water, but I cannot say that I have been satisfied with the results obtained by this use of the drug.

Bad cases require rest in bed, particularly anæmic girls and women.

The stomach will have a better opportunity to get well when at rest than when at work; thus it becomes necessary, sometimes, to abstain from feeding by the mouth altogether. Rectal alimentation then comes in to great advantage, for in conditions of such genuine starvation, the lymphatics are very greedy, and absorption from the rectum is very active; but the rectum *absorbs only*, and *does not digest*, thus whatever has to be absorbed must have been digested previously—peptonized.

Ulcer of the stomach, in both the young and old, being frequently associated with intense anæmia, the result, in these as in many other cases, is mistaken for the cause. The iron, the great presumed panacea for anæmia, is introduced into the stomach *which cannot digest it*, and in its attempts to do so, pain, ulceration, and danger are increased; particularly is this case with the chloride of iron, which is, more than any other, a vascular irritant. The lactate and malate are better preparations; also the subcarbonate of iron with sugar, of the Pharmcopœia. But if everything be correct which has been said in its favor by Gerhardt, Eulenburg, Uffelmann, and others, all these preparations appear to be excelled by the liquor ferri albuminati.

GASTROMALACIA. SOFTENING OF THE STOMACH.

It has always been a question whether softening of the stomach should be regarded as a vital or a cadaveric process; at all events, there are no symptoms which belong to it as such. It consists in a pultaceous breaking down of the tissues, mostly in the fundus. and is chiefly found after either exhausting diseases, or those connected with excessive gastric fermentation going on before and after death. It is true that it is now and then observed where it appears to have existed during life—for instance, in tubercular meningitis. It has also been claimed that, in these cases, it may depend upon, or be connected with, the high temperature of the patient during the last few days of the disease. But the high temperature of pneumonia, of scarlet fever, or intermittent fever does not produce it, and it is probable that in cases of tubercular meningitis it is the result of the insufficient circulation and defective nutrition during the exhausting disease, and the trophic changes taking place under the influence of nervous exhaustion.

Thorspecken reports a case of a baby three

months old, suffering from sleepiness, cough, dyspnœa and diarrhœa. All at once there was nausea, and an audible snap, followed by vomiting of bloody liquid and death within two minutes. In the fundus of the stomach there was an opening two centimetres in diameter, surrounded by softened tissues, and accompanied by tubercles in the lungs and spleen.

This case would be more conclusive if it were more intelligible. We do not know whether the perforation was due to the softening that takes place during the dying process, or as in the above cases of tubercular meningitis, or whether it was due perhaps, to tubercular ulceration. At all events, it appears improbable that the perforation of softened tissue would be attended with any audible snap.

Why is it that gastromalacia is found much more frequently in the bodies of little babes than in advanced age? It is probable that it is due to the fact that more acid is secreted, and the secretion of acid is continued for a longer period, wellnigh to the end.

Elsæsser wrote a monograph on this subject in 1846; it is still the best and most conclusive. He arrives at the conclusion that gastromalacia is a cadaveric occurrence, for the following reasons:

First.—It occurs mostly in the lower part of the stomach.

Second.—A healthy stomach removed from the body softens under the influence of acid fermentation.

Third.—The acid and fermenting contents of the softened stomach, introduced into the gastric organs of healthy animals, softened their stomachs, as was proved by autopsy when they were speedily killed.

Fourth.—There is no vital reaction around the softened spots.

TUMORS OF THE STOMACH.

Tumors of the stomach are very rare in infancy and childhood. Several varieties, however, have been observed, such as lipoma, fibroma, myoma, myo-sarcoma, sarcoma, adenoma, dermoid cysts, and multilocular lymphangioma.

Albers met with a cyst two and a half inches long on the small curvature of the stomach of a child.

Congenital carcinomata are very rare, but have been observed. Scheffer has described a case of encephaloid cancer which spread from the stomach to the spleen of a child twelve years old. In that case there was, besides the presence of a tumor, pain, vomiting, emaciation and collapse.

No doubt, heredity exerts its influence, but it certainly does not exist in every case. For, Hauser appears to have proved positively that cancer may result from epithelial proliferation originating in the cicatrices of ulcers. Carcinoma begins in the mucous membrane; the form called cylindroma in the glands.

As a common symptom hydro-chloric acid is absent from the secretions of the stomach in carcinoma.

Tubercles and tubercular ulcerations are met with in connection with general miliary tuberculosis at an early age. Caron reports the case of a girl twelve years old, who suffered from pain and bloody vomiting. Otherwise in these cases there are scarcely any symptoms, except the general ones belonging to the disordered stomach and the tuberculosis.

Pepsin in Infantile Diarrhœa.

Statistics show that the mortality rate of infantile diarrhœa, as it manifests itself in the summer months, is higher than that of any other disease.

How shall the conditions present best be met? To answer this query has inspired exhaustive contributions from the pens of our most learned medical writers. It is admitted by all that one of the causes which incites and perpetuates the gastric and intestinal inflammation is undigested, or partly digested, fermenting milk or other food, the decomposition of which is accompanied by the development of ptomaines and other toxic principles. It is an aid to the removal of this cause, both in predigesting milk or other food before it is given, and in digesting fermented undigested food in the stomach, that pepsin is indicated in infantile diarrhœa, and its efficacy has been attested by many well-known medical writers. (See *J. Lewis Smith, M. D., Archives of Pediatrics, Sept., '86, p. 518; Nov. '86, p. 639; Nov. '64, p. 424. Prof. Vocher, of Berlin Archiv. f. Kinderh, vol. 9, p. 3. Dr. I. N. Love, St. Louis Weekly Medical Review, Aug., '88. T. Lauder Brunton, Diseases of Digestion, p. 291. A. Holt, N. Y. Archiv. Pediatrics, 1886, p. 732. A. G. Bigelow, Archiv. Pediatrics, 1884, p. 430. Discussion at German Medical Congress, at Salzburg, 1881, by Biedert, Gerhart, Henoch, Steffen, Thomas, Soltman, Pfeiffer. Prof. Leeds, Archiv. Ped., 1884, p. 421, etc.*)

With the improvements that have of late been made in the purity, quality and efficacy of Pepsin, this agent is likely to play a more important and definite part in the treatment of intestinal inflammations than ever before. Its ease of administration, its certainty of action when a proper product is administered, will, we believe, lead to its extensive use.

We guarantee the purity, activity and solubility of our pepsin products. Our pepsin is absolutely free from odor, and has been shown by expert examination to be free from ptomaines and leucomaines, and demonstrated by an exhaustive comparative test to possess twice the digestive power of the most active hitherto introduced. (See Observations on Digestive Ferments, by R. H. Chittenden, Ph. D., *Philadelphia Medical News*, February, 16, 1889).

We supply pepsin in the following forms:

Pepsin Purum in Lamellis; Pepsinum Purum Pulvis; Pepsin, Saccharated, U. S. P., 1880; Pepsin, Glycerole, Concentrated; Pepsin, Lactated; Pepsin Liquid, U. S. P., 1880; Pepsinum Purum Tablets, 1 gr., Sugar-Coated, Pepsin Cordial.

All information desired by physicians as to our pepsin products, our general line of standard medicinal preparations, pharmaceutical specialties, and latest therapeutic novelties and improvements in methods of medication, will be promptly furnished on request.

PARKE, DAVIS & CO.,
DETROIT AND NEW YORK.

IN EXPLANATION
OF
The Physicians' Leisure Library.

We have made a new departure in the publication of medical books. As you no doubt know, many of the large treatises published, which sell for four or five or more dollars, contain much irrelevant matter of no practical value to the physician, and their high price makes it often impossible for the average practitioner to purchase anything like a complete library.

Believing that short practical treatises, prepared by well known authors, containing the gist of what they had to say regarding the treatment of diseases commonly met with, and of which they had made a special study, sold at a small price, would be welcomed by the majority of the profession, we have arranged for the publication of such a series, calling it **The Physicians' Leisure Library.**

This series has met with the approval and appreciation of the medical profession, and we shall continue to issue in it books by eminent authors of this country and Europe, covering the best modern treatment of prevalent diseases.

The series will certainly afford practitioners and students an opportunity never before presented for obtaining a working library of books by the best authors at a price which places them within the reach of all. The books are amply illustrated, and issued in attractive form.

They may be had bound, either in durable paper covers at **25 Cts.** per copy, or in cloth at **50 Cts.** per copy. Complete series of 12 books in sets as announced, at **$2.50**, in paper, or cloth at **$5.00**, postage prepaid. See complete list.

PHYSICIANS' LEISURE LIBRARY

PRICE: PAPER, 25 CTS. PER COPY, $2.50 PER SET; CLOTH, 50 CTS. PER COPY, $5.00 PER SET.

SERIES I.

Inhalers, Inhalations and Inhalants.
 By Beverley Robinson, M. D.
The Use of Electricity in the Removal of Superfluous Hair and the Treatment of Various Facial Blemishes.
 By Geo. Henry Fox, M. D.
New Medications, in 2 Vols.
 By Dujardin-Beaumetz, M. D.
The Modern Treatment of Ear Diseases.
 By Samuel Sexton, M. D.
The Modern Treatment of Eczema.
 By Henry G. Piffard, M. D.
Antiseptic Midwifery.
 By Henry J. Garrigues, M. D.

On the Determination of the Necessity for Wearing Glasses.
 By D. B. St. John Roosa, M. D.
The Physiological, Pathological and Therapeutic Effects of Compressed Air.
 By Andrew H. Smith, M. D.
Granular Lids and Contagious Ophthalmia.
 By W. F. Mittendorf, M. D.
Practical Bacteriology.
 By Thomas E. Satterthwaite, M. D.
Pregnancy, Parturition, the Puerperal State and their Complications.
 By Paul F. Mundé, M. D.

SERIES II.

The Diagnosis and Treatment of Haemorrhoids.
By Chas. B. Kelsey, M. D.

Diseases of the Heart, in 2 Vols.
By Dujardin-Beaumetz, M. D.
Translated by E. P. Hurd, M. D.

The Modern Treatment of Diarrhoea and Dysentery.
By A. B. Palmer, M. D.

Intestinal Diseases of Children, in 2 Vols.
By A. Jacobi, M. D.

The Modern Treatment of Headaches.
By Allan McLane Hamilton, M. D.

The Modern Treatment of Pleurisy and Pneumonia.
By G. M. Garland, M. D.

Diseases of the Male Urethra.
By Fessenden N. Otis, M. D.

The Disorders of Menstruation.
By Edward W. Jenks, M. D.

The Infectious Diseases. In 2 vols.
By Karl Liebermeister.
Translated by E. P. Hurd, M. D.

SERIES III.

Abdominal Surgery
By Hal C. Wyman, M. D.

Diseases of the Liver.
By Dujardin-Beaumetz, M.D.

Hysteria and Epilepsy.
By J. Leonard Corning, M. D.

Diseases of the Kidney.
By Dujardin-Beaumetz, M. D.

The Theory and Practice of the Ophthalmoscope.
By J. Herbert Claiborne, Jr., M. D.

Modern Treatment of Bright's Disease.
By Alfred L. Loomis, M. D.

Clinical Lectures on Certain Diseases of Nervous System.
By Prof. J. M. Charcot, M. D.

The Radical Cure of Hernia.
By Henry O. Marcy, A. M., M. D., L. L. D.

Spinal Irritation.
By William A Hammond, M. D.

Dyspepsia.
By Frank Woodbury, M. D.

The Treatment of the Morphia Habit.
By Erlenmeyer.

The Etiologly, Diagnosis and Therapy of Tuberculosis.
By Prof. H. von Ziemssen.
Translated by D. J. Doherty, M. D.

SERIES IV.

Nervous Syphilis.
By H. C. Wood, M. D.

Education and Culture as correlated to the Health and Diseases of Women.
By A. J. C. Skene, M. D.

Diabetes.
By A. H. Smith, M. D.

A Treatise on Fractures.
By Armand Després, M. D.

Some Major and Minor Fallacies concerning Syphilis.
By E. L. Keyes, M. D.

Hypodermic Medication.
By Bourneville and Bricon.

Practical Points in the Management of Diseases of Children
By I. N. Love, M. D.

Neuralgia.
By E. P. Hurd, M. D.

Rheumatism and Gout.
By F. Leroy Satterlee, M. D.

Electricity, its application in Medicine.
By Wellington Adams, M. D.

Taking Cold.
By F. H. Bosworth, M. D.

Auscultation and Percussion.
By Frederick C. Shattuck, M. D.

Series IV will be issued one a month, beginning with November, 1889.

GEORGE S. DAVIS, Publisher,

P. O. Box 470.   Detroit, Mich.

www.ingramcontent.com/pod-product-compliance
Lightning Source LLC
Chambersburg PA
CBHW030340170426
43202CB00010B/1191